BECOMING TIMEFARING

TIME TRAVEL & THE HUMAN FUTURE

ERIC WARGO

Art by David Metcalfe

Cup+Saucer Press
Fairfax, Virginia

An Original Publication of Cup+Saucer Press
Becoming Timefaring: Time Travel & the Human Future

ISBN: 979-8-9924052-0-0

Book design: Eric Wargo
Cover illustration: “Saucer City” by David Metcalfe
Interior art by David Metcalfe

The author can be contacted via his website:
ericwargo.com

Cup+Saucer Press
Fairfax, Virginia

Contents

For my timeship pilot, Emily Wargo, and my logbook keeper, May Wargo

Introduction

Exploration is unthinkable without imagination. An explorer must have in mind an image of where they are going and how they are going to get there. But those imaginings don't have to be accurate. They can be vague. They can be wrong. They often prove wrong.

For the Polynesian navigators who made long voyages across the Pacific in prehistory, there was always something over the horizon, a There they could visit and, Maui willing, come back from, even if they didn't always know what it would look like. They didn't set off in the expectation that they might fall off the edge of a flat earth or go into an empty formless void. The same was true for the Norse explorer Leif Erikson, whose longship reached Newfoundland sometime around year 1000 of the Common Era. The forested land that Leif and his kin found and wintered in (Vinland) might or might not have matched his imaginings*—it certainly was different from the colder Greenland of his upbringing. And of course, as Americans learned in school as part of our national mythology, the Genoese explorer Christopher Columbus thought his ships were bound for the East Indies when he set off across the Atlantic in 1492; he landed in the Bahamas instead. He was wrong in all kinds of ways, of course—but he trusted he was headed *some*where. He had an imagination. Explorers always do.

For more than a century, futurists have dreamed of space exploration and, more recently, of putting colonies on Mars and exploiting the solar system's material resources. Those are potentially achievable technological dreams, and science-fiction writers have given us plenty to go on, imagination-wise. Today's private space ventures like SpaceX, realistically or not, aim to make us a multiplanet species using vessels that are straight out of the mid-20th-century novels of Robert A. Heinlein and Ray Bradbury. But I will be arguing in this book that the real final frontier, maybe even the next frontier, is going to be the fourth dimension, time. Our destiny, if we do not destroy ourselves, is to become timefarers, capitalizing on the informational as well as material riches of humanity's—and the universe's—history, no longer being confined to our narrow historical moment.

What do I mean by timefaring? I don't just mean vanilla travel into the future via relativistic spaceflight

---

* A previous explorer had already seen the coast of North America, according to *The Saga of the Greenlanders*, but did not make landfall.

and/or cryosleep, where an astronaut wakes up decades, centuries, or millennia after their departure. I mean leaving our normal temporal reference frame, traveling into the past, and ideally returning from that past back to the present. I also, and more immediately, mean interacting directly with our future on a two-way, not a one-way, basis via novel communications and computing technologies. In other words, real round-trip flights and real two-way communication across the fourth dimension. Realistically, if we hope to get farther into outer space than our own solar system, then new ways of manipulating the spacetime continuum will be needed, and those are tantamount to time travel too. Faster-than-light travel is travel into the past.

These are no longer just science-fictional ideas. Einstein's theory of relativity unlocked the first theoretical door early in the 20th century; but various conceptual challenges and even taboos acted as a doorstop, keeping the idea of time travel mostly unexamined, at least in mainstream physics theorizing. Quantum physics, which came hot on relativity's heels, unlocked other theoretical doors into that same exotic room—but again, those challenges and taboos kept people from really peering inside. That, as we will see, is starting to change.

The motives of traveling or communicating across time are myriad, as are the possibilities. Unfortunately, for reasons that will become clear, you can't go back and kill the baby Hitler or prevent the death of a loved one you are mourning—the first things many people imagine when the subject of time travel is raised. But there's a whole lot more you *can* do. The past holds limitless resources inaccessible in the present, as well as limitless information. And in the past, you can set the stage for present and future advantages that are otherwise unobtainable.

Optimistically, becoming a timefaring civilization could mean becoming something like history gardeners, benevolently shepherding our civilization through perils via deft societal stewardship, planting seeds in the past for a better tomorrow. Pessimistically, becoming timefaring could bring new forms of transtemporal colonialism and exploitation that are far more bizarre and scary than the intercontinental versions perpetrated by seafaring Europeans like Columbus and those who followed. As with nearly every technological achievement in human history, the reality will undoubtedly be some mix of positive and negative, and there is no telling how exactly things will unfold. We need to be prepared for the possibilities, though. We, here, now are living in the immediate prehistory of our timefaring future.

But Houston, we have an imagination problem, and oddly enough, it is our science-fiction writers who created it. Those Bradburys and Heinleins and Isaac Asimovs and Harlan Ellisons who gave us inspiring and useful visions of interplanetary realities utterly failed us in the task of imagining the past as a There that can realistically be visited or of the future as a real place with people and technology that we could already be in contact with. Instead, they sent us down myriad blind alleys, telling entertaining but often superficial stories about changing the timeline in some catastrophic way. And physicists who imbibed those stories as kids internalized their flawed assumptions.

Consequently, our imaginative maps of the fourth dimension remain filled with sea monsters around the edges that have the name "Paradox." Especially in recent years, time-travel storytellers in fiction and film evade the headache-causing complexities of truly four-dimensional thinking with plot devices like parallel worlds and alternate histories, a "choose your own adventure" picture of reality that is echoed in pop-metaphysics speculations about simulated realities and multiverses. It makes for fun escapism, but the far more useful challenge, I will argue, is to think in and with real, actual history, a history that is singular but sometimes loops back on itself in surprising and surreal ways.

Historically, whenever futures have arrived, they have been unevenly distributed. Aerospace industry historians have recently suggested that time travel, like the related topic of antigravity, has been at least a peripheral area of secret physics thinking, and possibly even an active area of research, since WWII.* And as James Gleick pointed out in his 2016 book *Time Travel*—albeit with insufficient curiosity—China banned its TV and movie scriptwriters from writing time-travel narratives in 2011.† Their excuse

---

* See Nick Cook, *The Hunt for Zero Point* (New York: Crown, 2003); Paul Schatzkin, *The Man Who Mastered Gravity* ([no city]: Paul Schatzkin, 2003).

† James Gleick, *Time Travel* (New York: Pantheon, 2016). The source for this is: Richard Brody, "China Bans Time Travel," *The New Yorker*, April 8, 2011, https://www.newyorker.com/culture/richard-brody/china-bans-time-travel.

was that such stories were undercutting national historical pride by showing characters escaping to pre-Communist China. But when countries ban stories about some technological advance, as the U.S. effectively did with atomic energy and atomic weapons during the time of the Manhattan Project, there's the possibility that something very important and secret—not to mention, societally and geopolitically disruptive—*could* be afoot.

Even without going down the most conspiratorial rabbit holes here, some of the technologies that will arguably make *informational* time travel possible are already openly being developed in academic physics and computing departments at top-tier universities, and these technologies could be a stepping stone to more radical technological achievements.* *Retrocausation*, a word that appears only once, offhandedly, in Gleick's book, is an emerging paradigm that may prove to be the key to these technologies.

Gleick may have missed such signals because he doesn't believe in time travel. He mainly sides with time-travel skeptics like Stephen Hawking, who we will see was so disturbed by time travel that he invented a whole "conjecture" (on really little scientific or mathematical basis) to disallow it. Treating the topic mainly as a literary-cultural trope about regret, the desire to change the past, Gleick effectively does for his readers what China's ban did for media audiences in that country: keeps the subject

---

* Julia Mossbridge, "How to Avoid the Time Wars," Medium, March 22, 2021, https://jmossbridge.medium.com/how-to-avoid-the-time-wars-4d35d023f137.

outside the sphere of serious (which these days means scientific) contemplation. Although I've been a Gleick fan ever since his 1987 book *Chaos*, I think he seriously missed the boat (or, missed the time machine) on this topic.

Whether the time-travel singularity is imminent or still farther off—I admittedly don't have a crystal ball, here—our timefaring future could get very, very weird, potentially very suddenly. So, this book is about the ways we need to update our thinking, and upgrade our assumptions about ourselves and about time, to prepare for a future where humanity explores its past and cultivates its long future.

Synopsis

Before we develop the technology needed to build a quantum chronophone, before we find a source of negative-mass material to open and stabilize a wormhole, and before we 3D-print the spacetime-warping metamaterials needed to build a timeship, the first step in becoming timefaring is applying the toolkit of the literary critic or deconstructionist. We need to travel back in pop culture and dismantle some of those sci-fi tropes that Gleick describes, most notably the infamous "grandfather paradox" and its variants. In Chapter 1, we will see that time paradoxes and "causality violations" are really nothingburgers, bugbears of faulty reasoning. They arise from a failure to distinguish *changing the past* from influencing, affecting, or *causing* events in the past. The latter is perfectly possible; changing the past, however, is a logical absurdity. Spoiler alert: So is *changing the future*. There are semantic pitfalls galore with this topic, and this chapter will try to help navigate them.

And before we can physically travel in time, a range of technologies already on drawing boards will likely allow the transfer of information or data across the fourth dimension, including sending information into the past or receiving information from the future (those are both the same thing). Chapter 2 will explore the new informational territory we could be about to enter where technologies can "pre-spond" to future events and where quantum computers may allow users to communicate with each other—or themselves—at different times.* The societal implications of a world with chronophones and "c-mail" are enormous—they go way beyond getting stock tips from our older self. New inventions and innovations could actually bootstrap themselves into existence thanks to the causal loops produced by these technologies.

Chapter 3 gets to the meat of time travel with a discussion of the three major technologies that could allow physically venturing across the fourth dimension: wormholes, faster-than-light travel, and time machines—or what should really be called timeships (because they have to travel in the three spatial dimensions too). These timefaring modalities have been frequently misrepresented in science fiction and have been depicted in movies in ways that hinder, rather than help, our understanding, so we will do

---

* Readers of my previous books will know that I think the brain is a biological version of such a device, and that strange causal loops are already our reality. See Eric Wargo, *Time Loops* (Charlottesville, VA: Anomalist Books, 2018).

some more mythbusting here. And since the possible realities of these modes of timefaring dovetail interestingly with aspects of UFO sightings, it is in this chapter that we will address the emerging school of thought that the UFO phenomenon could represent visitation by our descendants, not beings from other planets.

How will chrononauts find their way in the timeline? This is a bigger question than it seems, because the fabric of spacetime does not come pre-engraved with time markers, nor is there any kind of universal clock that could help a timeship pilot find a particular historical destination. (It's no fair having your time machine come equipped with a console that lets you dial in a target like the Crucifixion or the D-Day Invasion—the common cop-out of TV scriptwriters.) Chapter 4 explores the question of temporal navigation and the various natural calendrical tools that could help chrononauts home in on a desired historical event or time period. Here, the field of archaeoastronomy—how ancient peoples reckoned calendrical time using variables of planetary motion and other natural processes—can be of use in our thinking.

Chapter 5 pulls back to a wide-angle view to consider the possibility, broached most famously by physicist John Wheeler, that intelligences in the far future might have a role in co-creating the universe, even influencing its birth, through something like backward causation. Wheeler didn't have actual physical time travel in mind, but the kinds of technologies our descendants might avail themselves of could have the deliberate or inadvertent effect of shaping or altering cosmic history or even (re)writing the natural laws established in the first moments after the Big Bang. Is the anthropic principle—the mysterious fine-tunedness of the universe for life and intelligence—a mere coincidence, or could it have to do with a future/past history of temporal meddling? I will suggest that a time-looping cosmos with advanced superintelligences manipulating nature on a fundamental level is a compelling alternative to current theorizing about reality as a simulation.

Finally, in Chapter 6, we examine perhaps the most profound implication of time travel in the self-consistent universe: the possibility of going back in time to meet your younger self (or conversely, of being met by your older self—again, two ways of describing the same event). Can we imagine such a thing and preserve our sanity, if not our sense of free will? What would you, either of you, say in such an encounter? Thinking about this problem can act a bit like a Zen koan, a riddle designed to unlock a kind of epiphany about the profundity of the actual, singular, self-consistent world we all coinhabit, versus the escapist multiverses and virtual worlds of recent popular physics and metaphysics.

I am a writer of speculative nonfiction, and there is a lot of speculation in this book—unashamedly. Unlike recent writers on the topic like Gleick, I would rather err on the side of being too fanciful (some of what you will read in these pages is obviously very fanciful) than rehearse the same caution and skepticism that scientists and science writers typically retreat to, perhaps to preserve their respectability. Life's too short to be respectable, or always right. The point here is to foster imagining. Our imaginings don't need to be accurate; they just need to be there.

This book should not be read as a comprehensive source on the technicalities of time travel. We will not get bogged down in debates about the exact quantity of negative mass needed for a warp drive, whether quantum fluctuations will make wormholes too unstable to traverse, and other questions that have occupied theoretical physicists. There are other good, popular books on the science of time travel that go into these kinds of technical details, and interested readers can find them cited in the footnotes.

The fact is, knowledge evolves. Science constantly makes new discoveries that render existing arguments, debates, and calculations obsolete. I very much agree with my friend and fellow time-researcher, neuroscientist Julia Mossbridge, that scientists in all fields, even brilliant physicists, can only see what their theories and assumptions allow them to see at any given time.* In the end, they can be just as wrong and shortsighted as the rest of us. This book follows the (current) science, but not slavishly.

Physicists can also be just as neurotic as the rest of us. The dramatic, conflictual, often violent scenarios found in physicists' thought experiments about time travelers trying to thwart or even murder their younger selves offer a fascinating window onto these neuroses, I think. Sigmund Freud believed that neuroses and taboos have a common origin, and as an anthropologist by training, I am partic-

---

* Mossbridge, "How to Avoid the Time Wars."

ularly interested in the taboos that have kept the subject of time travel and the related subject of backward (retro-) causation so marginalized and stigmatized for so long. Highlighting those taboos is as much a part of this book as the scientific speculation.

The time frontier is wide open to thinking and theorizing from many perspectives, not only those of physicists but also those of artists, of philosophers, of anthropologist-futurists like myself, and of ordinary people unburdened by too many academic assumptions, temporal neuroses, or the need for respectability.

So, without further ado, let's imagine…

1. Paradoxes, Assigned Paths, & Other Time Travel Fallacies

"And that," he said, "is the Path, laid by Time Safari for your use. It floats six inches above the earth. Doesn't touch so much as one grass blade, flower, or tree. It's an anti-gravity metal. Its purpose is to keep you from touching this world of the past in any way.

"Stay on the Path. Don't go off it. I repeat. Don't go off. For any reason! If you fall off, there's a penalty. And don't shoot any animal we don't okay."

"Why?" asked Eckels. ...

"Crushing certain plants could add up infinitesimally. A little error here would multiply in sixty million years, all out of proportion. ... Who knows? ... But until we do know for certain whether our messing around in Time can make a big roar or a little rustle in history, we're being careful. This Machine, this Path, your clothing and bodies, were sterilized, as you know, before the journey. We wear

these oxygen helmets so we can't introduce our bacteria into an ancient atmosphere."

"How do we know which animals to shoot?"

"They're marked with red paint," said Travis.[*]

This bit of dialogue comes from one of the great classics of the time-travel genre, Ray Bradbury's 1952 short story, "A Sound of Thunder." Sometime in the mid 21st century, safari hunters are able to pay big bucks to travel 65 million years into the past for the opportunity to bag the ultimate trophy, a T-rex. The dinosaurs serving as game for the hunters are pre-selected (and marked with red paint) by Time Safari Inc. because they were shortly going to die anyway of old age or some other accident, thereby minimizing the potential history-altering impact of this venture. And to keep clients from interacting in other ways with the local flora and fauna, the hunters must stay on an assigned antigravity Path the company has installed.

Of course, the company's prophylaxes against history-changing prove insufficient. When a group of hunters return to their present day, they find a subtly different world from the one they remembered: The words of a sign in the Time Safari, Inc. offices are spelled differently. Worse, a fascist has been elected president even though the travelers remembered that his democratic opponent had won the vote prior to their departure. One hapless hunter, Eckels, discovers a crushed butterfly on his muddy boot—

* Ray Bradbury, "A Sound of Thunder," retrieved January 11, 2025, https://www.astro.sunysb.edu/fwalter/AST389/ASoundofThunder.pdf.

he had ventured off the walkway briefly—and the group realize that that single crushed insect set history and human civilization on a slightly but importantly altered course. The "sound of thunder" Eckels hears at the end is the gunshot of his own swift execution for causing such a drastic historical disaster. Harsh punishment in the fascist future.

"A Sound of Thunder" is typical of many time-travel yarns in which history is presumed to be a delicate tissue, subject to alteration by the slightest interference. The altered timeline could be just slightly different (like differently pronounced words or a different election outcome) or it could be different enough that there is no time machine or time traveler at all, in which case the journey to the past ought to be foreclosed.

The term *grandfather paradox* has its origins in a 1929 editorial by Hugo Gernsback commenting on a time-travel story by one of the contributors to his magazine, *Science Wonder Stories*. What, he imagined, if he "visit[ed] the homestead of my great great great grandfather, and I am able to take part in the life of his time. I am thus enabled to shoot him, while he is still a young man and as yet unmarried. From this it will be noted that I could have prevented my own birth ..."*—which of course would prevent himself from ever going back in time to commit this meta-patricide, and so on. Gernsback loved the

* Quoted in Paul Nahin, *Time Machines* (New York: Springer-Verlag, 1999), 254.

(im)possibilities, and solicited stories on this theme for future issues.

That it would be incumbent upon any time traveler to be very careful not to alter history became a widespread assumption that dominated this science-fiction sub-genre. In the absence of something like Bradbury's assigned antigravity walkway, visitors to the past must tiptoe, disturb as little as possible, and preferably, get out of there quick.

In Octavia Butler's 1979 novel *Kindred*, for example, a modern-day Black writer named Dana is repeatedly summoned through time to an antebellum Maryland plantation. There and then, she must endure the nightmarish reality and avoid harming the slaveowner's son—her multiple-great-grandfather—long enough to ensure that her ancestor Hagar is born from the rape of one of his slaves. (Based on this novel, activist and time historian Rasheedah Phillips has coined the term "Black Grandmother Paradox" to centralize, among other things, the helping role of a time traveler toward their past kin, versus the murderous assumptions that often prevail in such narratives.[*])

Preventing the past from being catastrophically altered was also the premise of Robert Zemeckis' *Back to the Future*. Marty McFly takes a souped-up time DeLorean back to 1955, where he must ensure that his high-school-age parents fall in love, etc. And the same assumption—

---

[*] Rasheedah Phillips, "[Black] Grandmother-Paradoxes," *BlackWomenTemporal*, accessed on January 19, 2024, https://www.blackwomentemporal.net/black-grandmother-paradoxes.

don't change the past—shows up at the end of the recent film, *Indiana Jones and the Dial of Destiny*: Indy's companion Helena prevents our hero from remaining in 212 BC Syracuse, where a time fissure has brought them, because he might "muck up" history if he stays too long. Spoiler alert: If you don't like spoilers, don't engage in time travel or even read articles about the subject, because sometimes the alerts come too late.

In these kinds of stories, there is always an ambiguity about whether history would really have been mucked up had the character not tiptoed, or if their visit to the past was already part of a self-consistent universe—more on which, shortly. But in some stories, the time traveler definitely does wind up mucking things up, and then they must exert great effort to un-muck them. That is the premise of the beloved 1967 *Star Trek* episode "The City on the Edge of Forever." In this Hugo Award-winning episode from the series' first season, based on an original teleplay by Butler's mentor and friend, Harlan Ellison, an accidentally drugged and deranged Dr. McCoy flees through a time portal in the heart of a ruined city on a dead world, and the landing party suddenly finds that the *Enterprise* is no longer in orbit overhead, so they are stranded. To make things right, Kirk and Spock have no choice but to go through the portal themselves to undo or prevent whatever it was McCoy had done in Earth's past.*

---

* For more on Ellison and "City on the Edge of Forever," see my article, "Edges of Forever: Time Travel and the Stories that Might Have Been," *The Nightshirt*, February 21, 2019, https://www.thenightshirt.com/?p=4359.

Changing-the-past stories are usually full of logical holes that even a 10-year-old could pilot a space cruiser through. An astute pre-teen viewer of "City on the Edge of Forever" will note, for instance, that there is no way for Kirk, Spock, Uhura, and some redshirts to still be standing on the planet surface if there's no *Enterprise* that could have brought them to the planet in the first place. Similarly, if history has been altered in Bradbury's "A Sound of Thunder," the time travelers themselves are products of the same altered history that produced the fascist election-winner, so how would they know anything was different? If history has changed, that should include everything they themselves had ever learned and experienced prior to getting in the time machine. (When people speculate about some timeline-altering event as explanation for the "Mandela Effect"—things turning out to be not quite as we remembered them—it commits the same error, amputating our consciousness and memory from the entirety of our past history. We'll come back to the Mandela Effect in Chapter 5.)

The best that skilled writers can do with changing-the-past scenarios is paper-over the logical flaws in their premise and distract you with storytelling sleight of hand. But unfortunately, a century of these kinds of narratives has exerted a hypnotic effect. Some version of Bradbury's assigned Path that time travelers must not deviate from *or else* is a trope that has lodged so deeply in our brains that it acts as a kind of, well, *assigned path* in our own thinking about the subject of time travel and its alleged paradoxes. It has left us with a flat-Earth view of time, affecting even our best minds. Spoiler alert: That includes famous physicists.

Peepholes and Wormholes

Our quarry is the fourth dimension, viewed from a world that only feels comfortable with three. To help cut through the time thicket, let me tell a true story about a world that was mainly comfortable with two dimensions and was unused to dealing with a third. In this world, even the most cutting-edge scientific geniuses could not quite wrap their heads around the possibilities and were utterly wrong in their thinking—for decades.

In the early 1400s, in Florence, there lived a brilliant young sculptor, and later architect, named Filippo Brunelleschi. He had an engineering mindset like many of his artist contemporaries, and the laws governing vision particularly interested him. From observing the palazzos and vistas in his town, he grasped how parallel lines recede to a point on the horizon, and he realized that a convincing illusion of spatial depth could be produced by recreating this effect on a flat surface through careful use of geometry.

In 1413, Brunelleschi painted a picture of a familiar building in his town demonstrating this new principle. It was a wooden panel roughly the size of a vinyl record sleeve, and on it was a depiction of the octagonal baptistery associated with the city's cathedral (for which he later, and more famously, designed an enormous dome). Although other techniques such as occlusion and making more distant objects smaller have been used for indicating spatial depth probably since the very beginnings of painting,

Brunelleschi's picture of the Baptistery was the first Western image we know of that was rendered in precise, geometric perspective, all parallel lines receding to vanishing points on an occluded horizon. Instead of painting the gray-blue sky above the baptistery, Brunelleschi applied a reflective silver foil to reflect the real sky, enhancing the viewer's illusion that they were looking at something real.

The panel no longer exists, and we have only various written recollections by Brunelleschi's contemporaries, attesting to its awe-inspiring, even life-altering impact on them. The most detailed account is that of his biographer, Antonio Manetti, who wrote that seeing the Baptistery this way was something truly special, an experience to be repeated: "To look at it, under the various specified circumstances … it seemed that one was seeing truth itself; and I held it in my hands and saw it several times in my own day, and so can testify to it."*

Brunelleschi's demonstration was really more a contraption than a painting, because it was not meant to be viewed straight-on from in front: Instead, it had a tiny hole drilled through the exact center of the panel (corresponding to a central vanishing point). The viewer was made to hold the panel up to his face, almost like a mask, and look through the peephole from the back, at a mirror held out in front of the panel at arm's length.

---

* In Hubert Damisch, *The Origin of Perspective* (Cambridge, MA: The MIT Press, 1995), 116.

The reasons for this convoluted requirement are the reason I am telling this story. Brunelleschi seems to have believed—he just assumed—that to see a 3D illusion in a flat perspectival rendering accurately, the viewer's eye had to be fixed, motionless, and to occupy exactly a very precise point in space directly opposite the picture's vanishing point—called the "station point."

No less a mind than Leonardo da Vinci asserted this same presupposition confidently in several places in his diary, when addressing the hot new subject of perspective. In one passage widely cited by art historians, Leonardo wrote that "since many (men's) eyes endeavor at the same time to see one and the same picture produced by this artifice, *only one can see clearly the effect of this perspective, and all others will see confusion*" (my italics).* Everyone viewing Brunelleschi's demonstration, including Manetti, would have assumed the same thing. Indeed, the peephole in the painter's contraption was not only to position the viewer correctly at the station point, but to ensure that only one person's eyes—or rather, eye—would see the panel at a time, lest multiple eyes jostle around it and see only confusion.

I hope I don't need to tell you that what Brunelleschi took for granted, and what Leonardo asserted confidently in his diary, is not true. Practically any time you hold a photograph in your hand or see one in a book or magazine, or any time you view pictures on the screen of a

---

* Quoted in Lew Andrews, *Story and Space in Renaissance Art* (Cambridge, UK: Cambridge University Press, 1995), 46.

smartphone or a computer, and for that matter any time you watch a movie in a theater, you are viewing perspective pictures. Photographs and cinema are all "in perspective" even more precisely than Brunelleschi's painted simulation could have hoped to be—and you see the depth in the picture just fine. You do not see confusion, nor is the experience much enhanced when viewing exactly straight-on, with a single eye, through a peephole, in solitude.

We don't even take perspective as an illusion anymore, it's such a taken-for-granted part of our visual culture. Yet 600 years ago in Italy, at the dawn of perspective, no less a genius than Leonardo, the greatest scientific as well as artistic intellect of his time or perhaps any time, believed in the impossibility of experiencing a realistic rendering of spatial depth other than in very precisely controlled circumstances, through a peephole.

One of two big takeaways of this story is that when there's some big conceptual revolution or breakthrough, a major paradigm shift, the initial impulse is to confine or constrain the new idea, to assume it only applies in very special or restricted circumstances and is either fragile or dangerous when applied more broadly. This is one way to look at all those stories about time travel with some permutation of that assigned Path imagined by Bradbury. It takes a while for people to get comfortable enough with the new paradigm to venture off their conceptual walkways, to see that their previously assumed restrictions don't really hold and that the new principle might be far more widely applicable than in the tiny confines of some preliminary experiment or demonstration.

The other big takeaway is that geniuses are sometimes wrong. Their contemporaries, being in awe of them, sometimes don't even think to question their pronouncements.

Today, still really at the advent of thinking in and with the fourth dimension, we encounter the same kinds of basic errors in reasoning by the voices dominating the scientific conversation that prevailed in Leonardo's day. No less a physicist than Stephen Hawking, for instance, argued that the possibility of grandfather-type paradoxes probably ruled out time travel, despite such travel being fully allowed by Einsteinian physics. It's as if these fellows have been looking at the problem through a narrow peephole.

Around the Block

Here's a good place to introduce a concept we will continually be referring to throughout this book. After Einstein published his first paper on relativity in 1905, his former math teacher Hermann Minkowski discerned one of its most striking and important implications: Instead of time moving like a river, as it is often conceived and as people do experience the flow of events subjectively, time is really a dimension, no different from the three dimensions of space. Minkowski called our four-dimensional reality a *space-time continuum*, but it has often assumed the shorthand term "block universe." (You will sometimes also hear the terms "block world" or "block time."*)

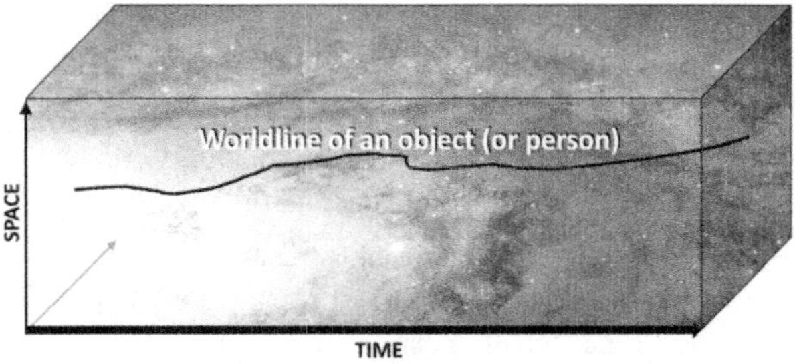

Block Spacetime of Einstein and Minkowski.
Past, present, and future coexist; objects are really "worldlines"

Most disturbingly to our everyday intuitions about free will, the block universe means that events in the future already exist, the same way events in the past do. What may seem to you right now like an open-ended future subject to your influence is already "in the past" of observers elsewhere, and thus thinking about your future as truly open-ended is as mistaken as thinking of the past as open-ended.

Dynamism or motion, too, is really an illusion in block world: Every object—every particle, person, or planet—is a "worldline" snaking through the block from past to future. What we perceive at any instant is really a cross-

---

* Nahin notes in *Time Machines* that describing time as a "block" actually preceded Einstein and even H. G. Wells, first appearing in Francis Herbert Bradley's 1883 *Principles of Logic*.

section of that 4D reality. Because we lack a suitable language to describe events and influences in block world, I will continue to use ordinary motion verbs like "travel" and "cause" (and even "retro-cause"), but you should imagine those words with an asterisk. Nothing—not time, not causation—really "flows," despite how fluid that line may appear.

Again, this is fundamentally at odds with how we perceive or experience time, as something moving from a fixed past toward an uncertain and yet-unmade future. One way of thinking about consciousness is as the cursor in a video-editing program moving along a timeline, or the moving bar of a scanner or copy machine, left to right. The perception of time's movement may in fact be a cognitive illusion, not unlike many perceptual illusions you may have seen in magazines or on the internet, where a particular arrangement of solid shapes seems to be in motion, and your brain can't unsee the motion even though you "know" nothing is really moving. Einstein hinted at this in a letter to the family of his close friend Michael Besso after Besso's death: that "the distinction between past, present and future is only a stubbornly persistent illusion."*

Physicists have grown accustomed to the illusoriness of past, present, and future, but many of them, and certainly many of their philosopher colleagues and people in the

---

* Douglas Clark, "Quoting Einstein: Einstein and Michele Besso," *Quoting Einstein*, June 27, 2013. http://quotingeinstein.blogspot.com/2013/06/einstein-and-michele-besso.html.

wider public, have found the challenge to free will presented by Minkowski's block universe too upsetting. Even Einstein was uncomfortable with certain implications of his theory, at first. But in the years and decades after he made his initial breakthroughs, experimental science has only validated his spatialized, relativistic conception of time. He was right—at least, scientifically right—even if we do not necessarily like it.

Einstein's friend, the mathematician Kurt Gödel, was actually excited by the new "eternalist" worldview that relativity pointed to. It was specifically to legitimize a block-universe vision of time that Gödel solved Einstein's field equations in a way that showed the possibility of a person actually traveling into the past and encountering their younger self. There are any number of paths through Gödel spacetime called *closed-timelike curves* that would allow such a self-encounter. Later physicists figured out that such paths could also exist in the vicinity of large rotating black holes, so those have served as a means of rapid spacetime locomotion for some science-fiction writers.

Nevertheless, perhaps partly because the topic had acquired a stigma from so much abuse in pulp science fiction, the possibilities of time travel sat on physics' back burner for decades. But then in the 1980s, physicists began to dust off their equations. It was really celebrity astronomer Carl Sagan who got the ball rolling. When he was writing his novel *Contact* in 1985, he needed a way to get his heroine Ellie Arroway (later played by Jodie Foster in Robert Zemeckis' 1997 film adaptation) to the Vega star system and back within her lifetime. Wanting to cleave as much as possible to real science, the astronomer reached out to a physicist named Kip Thorne at Caltech, who was the leading expert on black holes. Sagan wanted to know if those exotic objects might make the journey he had in mind possible.

According to Einsteinian physics, gravity is the curvature of space into the time dimension that occurs in the vicinity of massive objects. It has been proposed that black holes, which radically bend space, could be bridges to other parts of the universe—falling into one, if you survived, you might emerge elsewhere from a "white hole," although this was (and is) very speculative. Thorne knew of a more suitable object for Sagan's needs, though. With a sufficient amount of a still-hypothetical exotic material having negative mass, and formed in just the right shape, you can theoretically warp spacetime enough to actually create a traversable passage between distant regions. Such passages were first proposed as early as 1916 by a Viennese physicist, Ludwig Flamm, but the possibility was rediscovered by Einstein and a colleague, Nathan Rosen, in 1935, so they are sometimes called *Einstein-Rosen bridges*. Einstein-Rosen bridge is a mouthful, so in the 1950s the physicist John Wheeler gave these cosmological exotica the name "wormholes," because they reminded him of subterranean burrows linking separate patches of ground.

The thing with wormholes is that you can use them for time travel, not just space travel. If you build a wormhole in the vicinity of a black hole, you can then drag one of the two mouths in toward the black hole and then back out again. Because of relativistic effects (time passes slower in a deep gravity well), the two mouths have become separated in time, so you now effectively have a time machine. An astronaut entering the mouth that stayed put will have exited the other mouth earlier and can have met her younger self, before her journey. As we are about to see, this is a normal and unproblematic situation for any astronaut—er, chrononaut—who is in a position to utilize such a setup, but it is a really head-scratching situation for us pre-timefaring types.

Thorne worked out the mathematics of wormholes for Sagan, and he and two coauthors published a seminal paper, "Wormholes, Time Machines, and the Weak Energy Condition," in *Physical Review Letters*.* This set off a flurry of papers in the late 1980s and early 1990s by physicists weighing in on the possibilities and impossibilities, a story well told in David Toomey's 2007 book *The New Time Travelers*.† Some were really disturbed by the idea of wormhole time machines. In a 1992 article in *Physical Review D*, Hawking proposed a "chronology protection conjecture" that protects history from time travel, to prevent the causality violations that might arise.‡ The only evidence he provided for his conjecture had to do with technical quibbles, including the debatable idea that

---

* Michael S. Morris, Kip S. Thorne, Ulvi Yurtsever, "Wormholes, Time Machines, and the Weak Energy Condition," *Physical Review Letters* 61 (13) (September 26, 1988): 1446-9.

† David Toomey, *The New Time Travelers* (New York: W. W. Norton & Company, 2007).

‡ S. W. Hawking, "Chronology Protection Conjecture," *Physical Review D* 46, 603 – Published 15 July, 1992.

particles traveling between closely placed wormhole mouths might make the passage unstable. Mainly, he was just allergic to the notion of that self-encounter across time that Gödel had proven should be possible.

Time travel is to a physicist—or at least, to some male physicists—a bit like a mirror placed near a territorial cuttlefish in an aquarium: provoking incessant violent attack. The first thing that comes to mind is conflict and/or murder: The chrononaut from the future will at least do everything in their power to prevent their younger self from entering the wormhole; or more likely, they will try to *kill* their younger self. In a popular piece for the *Daily Mail*, Hawking offered the following thought experiment to illustrate how he saw the problem:

> I don't like the way scientists in movies are often described as mad, but in this case, it's true. This chap is determined to create a paradox, even if it costs him his life. Imagine, somehow, he's built a wormhole, a time tunnel that stretches just one minute into the past.
>
> Through the wormhole, the scientist can see himself as he was one minute ago. But what if our scientist uses the wormhole to shoot his earlier self? He's now dead. So who fired the shot? It's a paradox.

> It just doesn't make sense. It's the sort of situation that gives cosmologists nightmares.*

Hawking concludes from his little scenario:

> Any kind of time travel to the past through wormholes or any other method is probably impossible, otherwise paradoxes would occur. So sadly, it looks like time travel to the past is never going to happen. A disappointment for dinosaur hunters and a relief for historians.

Just like that, the famous author of *A Brief History of Time* wipes his hands of the messy business of time travel and dismisses it as "probably impossible" because otherwise we'd suddenly decide to shoot our younger selves through wormholes. Violent scenarios like this are not irrelevant to the cognitive distortions around the topic and mirror those older pulp writers' obsessions with grandfather-killing and dinosaur trophy hunting.

Hawking was by no means alone in fearing the sea monsters called "Paradox." In the 1960s, Australian theoretical physicist Brandon Carter asserted that the apparent time-travel possibilities opened up by certain kinds of black holes were "pathological."† Another physicist, the later Nobel-laureate Roger Penrose, agreed that the

---

* Stephen Hawking, "How to Build a Time Machine," *Daily Mail*, April 27, 2010, https://www.dailymail.co.uk/home/moslive/article-1269288/STEPHEN-HAWKING-How-build-time-machine.html

† Quoted in Nahin, *Time Machines*, 90.

implications of time travel in the vicinity of black holes meant that the whole theory that allowed them—relativity—might have to be abandoned. In his 1984 classic of spacetime speculation, *The Fourth Dimension*, science writer Rudy Rucker accedes to the mainstream physical consensus that time-travel experiments are problematic (if not impossible) because they could easily create paradoxes. He even speculates, "Perhaps some kind of 'Time Police' might be recruited to prevent such experiments."[*] (It's hard to tell which is more anthropocentric, the idea that the fabric of spacetime is so fragile that our puny experiments might cause irreparable harm to it, or that it could be up to human law enforcement to safeguard the universe from temporal damage.) Celebrity physicist Michio Kaku, in his 2006 book *Parallel Worlds*, recites similar objections and comes down in the "highly unlikely" camp, writing "even the tiniest disturbances into the past may cause unexpected paradoxes in the present" and "a genuine time paradox would be enough to completely rule out time travel."[†]

I could go on reciting the hysterical reactions produced by the topic of time travel in the physics community and in popular writings on the subject—Paul Nahin's 1998 book *Time Machines* is a wonderful resource on this strange byway in physics and its echoes in fiction. But as you probably by now suspect, these otherwise brilliant modern

---

[*] Rudy Rucker, *The Fourth Dimension* (Boston, MA: Houghton Mifflin Company, 1984), 171.

[†] Michio Kaku, *Parallel Worlds* (New York: Anchor Books, 2006), 144.

men are all just as wrong about time travel as Leonardo was about perspective in 15th-century Florence.

You Can't Get There from Here

When I was little, I used to play a game with reality, turning around really fast to see if I could catch it in the act of constituting itself (I couldn't). Sometimes I would also tiptoe to the threshold of my dark bedroom and suddenly flip on the light switch, trying to catch a brief glimpse of what the dark looks like (or whatever might have been hiding in it before scurrying away). Cautionary stories about changing history, including physicists' thought experiments about chrononauts and wormhole-builders changing the past, always rest on somebody, sometime, spontaneously deciding to do something new and unexpected, and history not waking up to it until after the fact, that moment of the individual's unexpected choice to go back in time and change things.

As Nahin points out, the inconsistencies and logical holes in most time-travel stories and thought experiments, including the idea of grandfather paradoxes, all arise from a simple mistake: failing to make the distinction between changing the past and affecting or influencing it. There's nothing paradoxical about a time traveler from the future doing things in their past that have effects, the same way all our actions in the present have effects: Our "freely willed" actions in the present influence what the future becomes, but they do not "change" the future from some *other* future that existed until we came along.* The latter notion is really nonsensical, although it lies at the heart of some of our cherished beliefs about agency, and it dovetails with the popular but problematic "many worlds" interpretation of quantum physics, as we'll see.

Thorne's 1988 paper inspired another physicist, Joseph Polchinski, to model the causal conundrums created by time travel using a less violent version of Hawking's thought experiment: shooting a billiard ball back one second in time through a wormhole, trying to deflect its younger self away from entering the wormhole, thereby producing a paradox. Billiard balls hitting each other are of course the classic symbol of cause and effect. But what Thorne and two of his students found, using linear algebra, was surprising—and reassuring. All possible paths of a billiard ball entering a wormhole later would, upon exiting the wormhole earlier, nudge its earlier self into the mouth of the wormhole later, thus completing the causal tautology, the closed-timelike curve.†

It seemed to confirm the thinking of a Russian friend of Thorne named Igor Novikov, who had long been fascinated by closed-timelike curves and who was clear-headed enough to not be deterred by the headache-causing weirdness that they produce. He had already determined on his own that a *principle of self-consistency* would prevent paradox in a universe that contained exotica like wormhole

---

* Nahin, *Time Machines*.

† See Fernando Echeverria, Gunnar Klinkhammer, Kip S. Thorne, "Billiard Balls in Wormhole Spacetimes with Closed Timelike Curves: Classical Theory," *Physical Review D* 44 (1991):1077-1099.

time machines, and thus he was terribly excited to see his American colleague's papers, as they gave him permission to write and talk publicly about this once-stigmatized subject.*

I prefer the term *time loop* to closed-timelike curve for describing such situations, because it is catchier. In some situations and scenarios we will discuss in this book, it is not spacetime per se that is looping but the chains of cause-and-effect that wend their way through it and sometimes circle back upon themselves; so in fact, *causal* loop would be a better catch-all term, but we'll mostly stick with time loop for consistency.

Whatever we call these causally circular formations, the implication of Novikov's self-consistency principle is profound: A time traveler will always find that whatever they do in the past, whatever mayhem they may freely commit, was already part of their own and/or their society's backstory before they took their trip. A universe that includes wormhole time machines (or, theoretically, any other mechanism of traveling in time) would require a greatly scaled up version of those equations about neurotic billiard balls, but the outcome would be no different. Killing your biological ancestors is off limits, not because the past can't be visited or your younger pre-you parents can't be met, but because in any scenario where you actually do visit your past, killing your ancestors didn't already happen. *It can't happen because it didn't.*

---

* Igor Novikov, *The River of Time* (Cambridge, UK: Cambridge University Press, 2004).

I like to tell people that if they are shopping for a good Zen koan, to short-circuit their old thinking and lead them to new insights (if not enlightenment), that last sentence is quite effective.

The universe cannot be fooled. History is not a fragile tissue, nor is it a senile grandfather unwary and unsuspecting of a young time traveler's dodgy spontaneity, their free will. Everybody's free will is already included in the march of time, already part of history, and accounted for in the larger scheme. No matter how quick or sneaky you are. But this is really hard to wrap our heads around, even if we are geniuses. Of all people who should have known better, even Hawking failed to distinguish affecting the past and changing it: In his chronology protection paper, he claimed that, even aside from his other arguments against time travel in principle, no one could venture to before the creation of a time machine because "that part of the history of the universe is already fixed"*—showing that he really thought travel to the past meant changing the past "from what it was."

A psychoanalyst might suggest that stories about dodgy time travelers and mad scientists taking history off guard are really a kind of projection. It is we fallible humans who are unsettled and surprised at the concept of time travel—it's still a pretty new idea, first arising in the Victorian era and ratified only in 1905 with Einstein's first paper on relativity, "On the Electrodynamics of Moving Bodies." Shocked and unsettled (also perhaps a bit

---

* In Toomey, *The New Time Travelers*, 329.

overexcited) by the possibilities, science-fiction writers and upset physicists sometimes concoct stories in which it is the universe that has been tricked or taken off guard, when it is really themselves who have been jostled loose from their familiar moorings.

The most interesting time travel narratives in science fiction take seriously this idea that history, no matter what, will be self-consistent, even in a universe of time travelers. Those are the stories we should pay attention to.

Consider for example Chris Marker's 1962 short film *La Jetée*, what I consider to be the best time-travel movie ever made. It begins on the "pier" (or concourse) at Orly airport, when a child witnesses a man being shot to death; the woman the man was approaching reacts in shock. Decades later, after a nuclear war has laid waste the earth, survivors living in Paris's sewers develop time-travel technology and seek to test it using prisoners. One of the prisoners is the man with the vivid pre-war memory of the assassination at Orly, and the experimenters determine that such a memory may be useful in stabilizing a time traveler's consciousness at the destination time. So they send him back to Paris before the war on a series of test missions; each time, he gravitates to the woman he remembered from Orly airport. They converse, take walks, visit a museum, fall in love. (Falling in love with a woman in the past is another common time travel trope—paging Dr. Freud.*)

The time traveler successfully completes his test missions and so he is then sent to the future to retrieve a power plant that can help civilization rebuild itself. Afterward, he realizes his captors will have no more use for him and will execute him, but time travelers from the far future intervene and help him escape back to prewar Paris, to find and live with the woman of his memories. He finds himself at Orly airport, the scene he remembered from his childhood, and runs to the woman—only realizing in the last instant that it was he himself whom he saw assassinated when he was a boy. His captors have pursued him through time and an assassin's gunshot prevents him from living in and with his fantasy.†

This story is a time loop: The man's memory of a woman watching a man being killed is what enables him, decades later, to be chosen for the time-travel experiment. His own ignorance of the role he played in that scene—as the man assassinated—is what enables him to play his fated role. You might mentally bookmark that word "ignorance," as we will keep coming back to it.

There are plenty of more recent examples, and they aren't always tragic, although they are always highly puzzling for the person or persons involved. For example, midway through Alfonso Cuaron's *Harry Potter and the Prisoner of Azkaban*, Hermione leads Harry several hours

---

* That "woman in one's past" is of course, at least symbolically, one's mother.

† *La Jetée* was the inspiration for Terry Gilliam's film *12 Monkeys*.

back in time using a "time turner" so they can rescue their slightly younger selves from a series of perils they had narrowly escaped in the first half of the movie. At each step of the way, they discover that various anomalies they had experienced the night before (a rock hitting Harry in the head out of nowhere, a sound that distracted a werewolf pursuer, etc.) were themselves trying to get their own attention or otherwise save their own lives. Hermione and Harry's journey into the past does not result in changing history—they are playing the very same role in the past that they evidently "already played"—but it results in their own greater understanding of events in the past that they had previously misunderstood.

We'll see more examples of this kind of scenario in the coming chapters. But bottom line: It is perfectly possible to tell stories about consistent histories in fictional worlds that contain time travel. And if human imagineers can do it, so, presumably, can the builders of the cosmos.

People trying to explain why time travelers can't change the past sometimes talk about "rules" preventing such things, but Novikov saw that these rules amount to nothing more or less than the constraints that also restrict our movement in the more familiar three dimensions of space. Many destinations in space are off limits, not because of some new physical principle, but simply because you physically lack the means or the wherewithal, or in some cases the needed backstory, to make the journey.

When I was picking his brain about the Novikov self-consistency principle for my book *Time Loops*, maverick quantum physicist/poet Nick Herbert (author of *Quantum*

Reality and other excellent books*) offered that, in physical terms, you could think of self-consistency as a kind of "force," analogous to the Exclusion Principle that keeps particles called Fermions (such as electrons) from occupying the same quantum states. Wolfgang Pauli won the Nobel Prize for this discovery in 1945. It is this force (although usually it is talked of as a "pressure") that gives us the periodic table of the elements: If you add an electron to an atom, it must occupy an already-empty space in one of the atom's shells/orbits or ascend to a new one. The same pressure prevents neutrons from condensing or degenerating below a certain point, which is what keeps a neutron star from completely collapsing under its weight into a black hole. You could even say that it is this exclusion principle or force that "holds the world up."

The rather mind-blowing (I think) implication of Novikov's self-consistency principle as a kind of force is that the past is "solid" the same way a spatial landscape is. There will be place-times in the past you can't readily get to, and things in the past you can't easily do and would not even think to do, just as there are lots of restrictions on my movements and behavior when hiking in the Colorado foothills, where I grew up. I can't just descend straight down into the solid dirt, for instance, or step through a rock outcrop or hillside. In my present bodily form, I can't soar into the air like an owl. There are also places where I would have to contort my body to squeeze through a narrow crevice or strain to climb a steep incline.

---

* Nick Herbert, *Quantum Reality* (New York: Knopf Doubleday, 1989).

There are also places you can't go, and many things you can't do, just because you lack the needed backstory. You can't go to the moon, for instance. It's not because moon travel is a paradox—people have been there. But *you* don't have a rocket. *You* didn't get straight A's in school and then train for decades to be a NASA astronaut. There's no moon rocket at Cape Canaveral right now that you could get in (or hijack), and so on. Despite the many degrees of freedom offered by being an animal able to move easily in your immediate environment, virtually all of the actual places that exist in the universe are off limits to you, now or in most cases ever, because there's no prior history putting you in a position to go there. *You can't get there from here.*

Past and future time, like those Colorado foothills with their pink sandstone formations, has a topography of possibility, solid contours in its landscape, defining what is accessible and thus subject to intervention and eliminating the need for assigned antigravity paths to "prevent paradox," as in Bradbury's story. The important thing to remember here is that I, a sometime scientist and alleged pseudoscientist, would have to be "mad" to even think about flying in the air, passing through solid stone or soil, etc. It would be like wanting to go to the moon and being frustrated that I (here, now, being who I am) cannot do that.

Situations in time-travel thought experiments labeled paradox, like scientists assassinating themselves through wormholes or time travelers killing their patriarchs, are really just places in the landscape of history that are impossible to get to from where they are, for exactly the same reason I can't go to the moon: The needed backstory is lacking. Remember your new koan: It can't happen because it didn't. *And if you think it did, you're wrong.*

Being Wrong

Being wrong about the past, not to mention all the myriad ways we are just ignorant about what happened in the past, turns out to be a huge part of the time-travel story. For instance, scenarios about time travelers trying to assassinate their grandfathers could theoretically arise, but only when they are extremely wrong—for instance, shooting the wrong guy because they were misled about who their real biological grandfather was (and thus, misled about how interesting their grand*mother* was). Same with time-portal makers trying to kill their younger selves.

Let's tweak Hawking's mad-scientist scenario slightly to see how it could be made to work. As he wrote it, the very premise is totally improbable, because anyone with the wherewithal (and real understanding) to build a wormhole would not harbor the primitive misunderstandings about time travel that Hawking and the pulp science-fiction writers of his youth harbored. That's partly what I mean by not having the needed backstory: It depends on the universe somehow changing its mind, or on the physicist tricking the universe, like child-me tricking reality with the light switch. But the scientist could perhaps become confused about the subject, or develop confused intentions, *as a direct result* of his meddling with history.

It might happen like this: Having built a wormhole time machine and engaging in final preparations for activating it for the first time so that he can *meet and shake hands with* his slightly older self, the quite sane and rational scientist sees a hole in space open up before him—precisely and excitingly as his calculations have predicted! His future self really did it! But as the shimmering portal stabilizes in the air, he sees something he did not expect: His slightly older self, bleeding from his head and with a crazed look, is pointing a pistol at him … and shoots!

The bullet from the future hits the scientist in the head … it's a small pistol, and his aim is poor because of the blood, so it's not a center shot and doesn't kill him immediately. (He's not a violent person, by nature, has never fired a gun before, hence his inaccuracy.) But the shot distorts his thinking and makes him confusedly wish to shoot himself in revenge, or perhaps self-defense—which is the only reason the gun was there to begin with. Thus, he reaches for the gun in the desk drawer a minute later as he flips the switch on his machine and opens his own wormhole to the past; he sees his wide-eyed, naively excited younger self, takes unsteady aim at the bastard despite blood flowing into one eye, and … you get the idea. In other words, it was his self-destructive temporal intervention that made the scientist "mad" in the first place.

Mad scientists are an effect, not a cause, of misguided thought experiments about changing the timeline.

I'm not kidding. In our timefaring future, there may be real mental-health implications of being wrong about time travel. An individual actually hijacking a time machine to their childhood or adolescence with the intent of undoing some mistake or preventing some tragedy would only be wreaking havoc on their own sanity. They would find themselves hopelessly confused by an experience in a time machine—sort of like putting a cat in a lunar lander, or a cuttlefish in a tank with a mirror—because of the misleading perceptions and false assumptions that led to their ever wanting to the change the past in the first place.

If you are paying attention, you will have noted something a bit disturbing about my revised Hawking scenario, and it might seem to you like a paradox: The scientist's self-annihilating act is a direct result of that self-annihilating act. Since he begins with only the best, friendliest intentions toward his older, er, younger self, there is *no* reason for shooting himself other than that he has already shot himself. It's an example of what is sometimes called the "bootstrap paradox"—something arising from the nothing of a time loop. But this is in fact the opposite of a paradox, a tautology—and those, I kid you not, are perfectly allowed in the universe.

In a 1992 paper on this theme, Novikov and a colleague, Andrei Lossev, called objects or information arising from the nothing of a time loop "Jinn," after the beings of Islamic folklore that sometimes just appear out of nowhere.* Savvy science-fiction writers and filmmakers have sometimes explored such Jinn in their stories. *La Jetée* has a bootstrap element: It is the time traveler's memories of a man being assassinated that enable him to travel in time. And James Cameron gave us a great model for it in his *Terminator* franchise. A future genocidal AI

---

* Andrei Lossev, Igor D. Novikov, "The Jinn of the Time Machine: Non-trivial Self-Consistent Solutions," *Classical and Quantum Gravity* 9 (October 1992): 2309-2321.

system called Skynet sends a robot assassin (Arnold Schwarzenegger) into the past in an unsuccessful attempt to kill the mother a future human resistance leader. The attempt fails, but fragments of the destroyed robot are retrieved by the defense contractor that will go on to develop Skynet by reverse-engineering those fragments. Where, in this causal loop, did the needed information to build Skynet come from? Skynet and its terminators are Jinn.

The bootstrap paradox is sometimes also called the "information paradox," and Michio Kaku cites it among his list of reasons why time travel is probably impossible:

> In this paradox, information comes from the future, which means that it may have no origin. For example, let's say a scientist creates a time machine and then goes back in time to give the secret of time travel to himself as a youth. The secret of time travel would have no origin, since the time machine the youthful scientist possesses was not created by him but was handed to him by his older self.*

Perhaps thumbing their nose at Kaku and other time-travel skeptics, the writers of the outstanding Netflix series *Dark* incorporated this exact scenario into their drama about a "time war" centered on a small German town. A clockmaker receives from a time traveler a copy of a book describing the principle of building a time machine, written by his future self. Based on this book, he builds the time

---

* Kaku, *Parallel Worlds*, 142.

machine … but again, there is no point in his life when he figures out the principles, for instance with equations on a chalkboard or some Eureka moment.

Bootstrap scenarios defy our understandings of causation and especially the arising of new knowledge, which we imagine must be hammered together from already-existing components in some linear cognitive way. But when you drill down, in a time-travel universe *everything* might in fact have a bit of self-causation in it. It is essentially the same thing as *creation ex nihilo*, which I argue elsewhere is the reality underlying artistic creativity and innovation.* The physicist John Wheeler even suggested that the universe itself may be self-caused, arising from an enormous, multibillion-year spacetime loop—a big Jinn, you might say. But the possibility of Jinn has proved so disturbing—traumatic, even—to some theorists that they have embraced alternative interpretations of physics entirely, such as multiverses. I will try to persuade you in the next chapter that multiverses are a conceptual cop-out, a way of escaping truths that just happen to cause headaches (if not actual head injury).

Homo Chrononauticus

The ocean explorer Jacques-Yves Cousteau once scandalized an audience of oceanographers at a London scientific congress by predicting that the next stage in human

---

* Eric Wargo, *From Nowhere* (Charlottesville, VA: Anomalist Books, 2024).

evolution would be humans surgically adapted to breathing seawater. Homo aquaticus, he said, would no longer require an aqualung but could live underwater indefinitely: "Homo aquaticus is completely at liberty in the sea: to play, to dream, to farm whales and herd fish, to repair undersea machinery, to supervise research."* The scientists in the room thought the Frenchman was nuts.†

Although you can still breathe ordinary air when visiting earlier or (hopefully) later time periods in human history, Homo *Chrononauticus* will be unlike us in some important, if subtle respects. They may not require surgical alteration so much as a psychological one, and it is possible that the modification to the species will be collective and societal, the result of a kind of cultural and even cognitive selection unfolding over decades, centuries, or longer.

First, to become timefaring or even just time-interacting (via something like the quantum servers we'll explore in the next chapter), it can no longer trouble people that they could possibly change history by their actions. The backstory of a timefaring civilization is one of clarity, not confusion, about time. Thinking the past can be changed is the product of a culture that disbelieves in time travel and does not understand it, and thus has not yet developed it. By the time the technology exists, a kind of selection will have filtered out history-altering intentions in the world of humans or posthumans who may find

---

* Jacques-Yves Cousteau, *The Ocean World of Jacques Cousteau, Volume 1: Oasis in Space* (Danbury, CT: The Danbury Press, 1973), 132.

† Brad Matsen, *Jacques Cousteau* (New York: Knopf, 2010).

themselves in a position to use that technology, and the very possibility of self-thwarting, changing the past, etc., will be a dim memory from a more superstitious, demon-haunted, pre-timefaring age.

In fact, as we will see, the self-consistent universe that includes time travel and causal loops is way stranger and more interesting than the flatland world bequeathed us by all those science-fiction writers and anxious physicists like Hawking with their tissue-like histories subject to damage and change. Our timefaring future (which, spoiler alert, is really our past) promises to be way more surreal—biblically surreal, even—than most people, myopically focused on the technical challenges and alleged paradoxes, suppose. But surreal is not the same thing as impossible.

Already in 1921, the physicist Hermann Weyl, who later was a colleague of Einstein and Gödel at the Institute for Advanced Study at Princeton, anticipated that the weirdness allowed by relativistic time loops resulted in "a spectral image of the world more fearful than anything the weird fantasy of E. T. A. Hoffmann has ever conjured up." But he added that "nowhere do we find any real contradiction to the facts ..."* In other words, *just deal with the strangeness*.

Today's touchstones in speculative fiction might be different than ghost story writer Hoffmann—Philip K. Dick, perhaps—but Weyl's sentiment is basically correct. Things will get very, very weird. Maybe they already are.

---

* Quoted in Nahin, *Time Machines*, 80.

2.
"You've Got C-mail!": Quantum Servers, Chronophones, & Other Modes of Transtemporal Communication

The story so far: Alterations to the historical timeline and associated "time-travel paradoxes" are non-problems arising from faulty reasoning, including the confusion that affecting the past is equivalent to changing it. Time travel is perfectly allowable in physics, and best of all, grandfathers are perfectly safe from their murderous timefaring grandchildren. It's not because causality police swoop in in flying saucers to prevent grandfather-assassinations, or because a paranormal force causes the gun to jam at the crucial moment. It's because there's simply no historical backstory for a mischievously paradox-causing misadventure in time. It can't happen because it didn't already happen.

Yet, all those fatal gunshots, echoing down through the generations of time-travel fiction and physicists' thought experiments, have had the effect of warning us off:

The past, if it can be visited, is therefore subject to alteration—so it probably can't be visited ... at least not safely. This has given us a flat-earth view of fourth dimension, and it has imposed a cognitive blockade on imagining traveling in it other than in the usual way—ploddingly, to grow old in a single direction along the one-way arrow of time. At best, we have been given comic-book fantasies of multiple forking timelines and alternative histories, a multiverse that lets us have the cake of time travel and eat it too.

One of the most interesting treatments of time travel in recent science fiction is William Gibson's 2014 novel *The Peripheral*. A gamer somewhere in the rural South in the mid 21st century named Flynne spends a few evenings helping her Marine vet brother beta-test a new "sim": piloting a quadcopter defending a glamorous inhabitant of a futuristic high-rise from a swarm of insectlike paparazzi drones. The ultimate objective or point of the game is never clear to her, nor is it clear to her brother, who was hired by a mysterious company to play it. It isn't just "some shooter"—and to Flynne, who is sick of game violence, it is weirdly satisfying just idly intercepting drones while observing the life and loves of the female character whose ultra-futuristic apartment she is defending. But on the second night of her play, she cannot protect the character from being gruesomely devoured by nanorobots that have managed to scale the building and get onto her balcony, seemingly with the help of the alluring woman's male companion.

Realities that turn out to be simulations are a staple of postmodern science fiction (as well as contemporary metaphysics, as we'll see in a later chapter), but this theme invites its antithesis: apparent simulations that turn out to be real. *Ender's Game* by Orson Scott Card is the most famous example of that premise, but *The Peripheral* takes it to a new, time- and mind-bending level: Disappointed at her failure to successfully defend her virtual target from attack, Flynne does not realize that the character whose apartment she was defending was (or, will be) real—the woman was not just some non-player character or NPC. She witnessed an actual murder, one taking place across the ocean, in London … and seven decades in her future. The victim has not even been born yet.

The Peripheral flickers back and forth between mid-21st century America and early-22nd Century London, where celebrity publicist Wilf Netherton seeks insights into the disappearance of the sister of one of his high-profile clients. In that future world, wealthy survivors of a late-21st-century global apocalypse called the Jackpot use an old quantum computer server in China (and dating to Flynne's time period) to hire gamers in the past for odd telepresence jobs, such as drone-piloting, without ever being witting to what they are really doing. The missing sister of Wilf's celeb client is the woman Flynne saw murdered on the balcony—so he uses the server to call Flynne, 70 years in the past, and find out what she saw. A future London police detective asks Wilf and his associates to send Flynne instructions for 3D-printing a virtual headset through which she can attend a celebration of the murdered woman's life, via a telepresence android (the "peripheral" of the title); there, she may be able to identify the man who let the killer-nanobots into the victim's flat.

(Making the headset is no sweat for Flynne, who works at the local 3D print shop.)

The idea of communicating across time via something like a quantum server is a fascinating premise, and as we'll see shortly, it could well be the real precursor to physical timefaring. But *The Peripheral* has one major flaw. Like many other time-travel storytellers, Gibson takes unnecessary pains to let himself off the paradox hook by having his characters live in divergent timelines once they have begun communicating across time. Wilf explains to Flynne that it is not really *her* future he is calling from and that her timeline is a "stub," branching off of his timeline.

Alternate realities or worlds have been a staple of fantastic literature (not to mention religion) since forever, but Murray Leinster's 1934 short story *Sideways in Time* is sometimes considered the first example of parallel historical timelines being used as a scientific plot device: A Virginia mathematician who had predicted a catastrophic timeline glitch leads his students through several alternative histories to try and get back to their own world. One of the most famous examples of the parallel-worlds genre is Philip K. Dick's 1962 novel *The Man in the High Castle*, set in an alternate historical timeline in which the Axis powers won World War II and have partitioned the former United States into two countries ruled by Germany and Japan, respectively. In the fractal manner of many of Dick's works, one of the characters in the story is the author of an alternative-history science-fiction novel in which the Allies won the war.

Writers frequently employ parallel timelines—or "continua" in the parlance of Gibson's *The Peripheral*—as a way to allow time travel without causing grandfather paradoxes. But if there are no grandfather paradoxes—if the whole idea is a bugbear of faulty reasoning—then there's no need for a "garden of forking paths" view of history (to use a metaphor from the Argentinean writer Jorge Luis Borges). I suggest that parallel worlds are another trope that needs to be jettisoned, like Ray Bradbury's assigned antigravity walkway, so that we can think rigorously about the problem of communicating or interacting across the fourth dimension.

The same self-consistency laws discovered by Igor Novikov for billiard balls traveling through wormholes will govern any kind of informational time travel, such as a message from a sender in the present to a recipient in the past. The sender will always find that however their message affected the past is already part of their own present backstory, which in most cases just means the diffuse background causal cloud of their society at whatever time point they happen to live. Gibson just didn't want to deal with these complexities, because as more of a cyberpunk writer, time's conundrums are not really his bag. But since there is no inconsistency in a self-consistent universe, there is no longer any traumatic rupture that would somehow bounce or deflect history onto a new track, nothing that would create a "stub."

Bathwater, Meet Baby

Gibson is most famous for his 1984 novel *Neuromancer*, which is widely considered to have been prophetic of the virtual worlds and cyberculture that took shape a couple decades later. Stubs notwithstanding, *The Peripheral* may be similarly prescient. An often-overlooked precursor to true, physical time travel using timeships or wormholes will likely be technologies that enable data transfer—communication as well as even virtual interactions—backwards and forwards across the fourth dimension. When this kind of *informational time travel* capability comes online, at scale, it could prove to be a decisive horizon for our civilization, technologically and societally.

As we have seen, time travel is really all about backstories. To understand how quantum servers might work and the effects these technologies could have, we need to rewind and first consider why people have so long disbelieved that anything of the sort could be possible. Some of the disbelief comes from those supposed time paradoxes we addressed in Chapter 1. But more fundamental (and perhaps part of the reason those tropes have proliferated) are longstanding scientific stigmas around the idea of any influence, such as a message, traveling from the future toward the past.

Information is just causation by another name; so when we are talking about sending a message back in time, we are talking about backwards causation.[*] And the idea that some future cause could exert an effect in the present, or *teleology*, is one that has been widely, almost universally, rejected by science ever since the Enlightenment. Although Aristotle had listed teleological (final) causes among the varieties of natural causation, the only

[*] Interested readers can find this elaborated in my book *Time Loops*.

teleology that Francis Bacon and Isaac Newton and other natural scientists in the 17th and 18th centuries could imagine had to do with God's divine plan. Since the new rule in science was that God couldn't be part of naturalistic explanations, teleology was thrown out along with Him.

This restriction worked fine for two centuries. Through the period when industrialism was ascendant, *efficient* causes—that is, causes "pushing" from the past—were sufficient to scientifically understand, predict, and control thermodynamic processes (i.e., processes involving heat and its transformations in work and its losses in entropy) and to build machines based on these processes.

But the Enlightenment's machine models of cause and effect began to break down at the turn of the 20th century, around the time that physicists were making major leaps in their understanding of electricity and radiation. It was discovered that at the smallest scales in nature—the realm of photons, atoms, and subatomic particles like electrons—there was an unpredictable component, which the astronomer-writer James Jeans likened to the "play" in an untightened mechanism.* The equations for electromagnetism are famously precise, but only for predicting large aggregates of particles. The behavior of an individual photon or electron can only be predicted by what is called a *wavefunction*, a hazy smear of probability. Even if you knew everything about the past history of an electron, you couldn't completely predict what it would do next. Even if

---

* James Jeans, *The Mysterious Universe* (Cambridge, UK: Cambridge University Press, 1930).

you knew everything that could be known about an atom of some unstable radioactive isotope, you would have no idea when it was going to decay—a second from now or thousands of years. This is the famous "uncertainty" that rules the domain of subatomic particles, and it was upsetting to scientists who had long believed that we live in a deterministic, mechanistic, therefore *predictable* universe, the universe of "classical" (versus quantum) physics.

Many explanations for uncertainty were proposed, and some argued that there had to be some not-yet-understood hidden variable that made things only appear to be random. As early as 1921, a German physicist named Walter Schottky proposed that a complete description of a physical system might need to include influences propagating backward from the future. However, the most forceful personality among the brilliant minds of the first generation of quantum physicists was the Dane, Niels Bohr, and he was not partial to the idea of hidden variables.

Bohr dominated the conversation at physicists' gatherings such as the decisive Fifth Solvay Conference in Brussels in 1927, and it was his "Copenhagen interpretation" that won out against rival interpretations of quantum weirdness. We have to just accept that randomness is an intrinsic property of nature at the smallest scales, he insisted. Only when a particle is subject to a final measurement (or "observation") does something definite emerge from that probability cloud. The birth of the actual out of quantum uncertainty, midwifed by observation, came to be called the "collapse of the wavefunction."

It is from Bohr that we get the evocative but misleading quantum-physical notion that observers create reality through their measurements. It probably does *not* mean that consciousness per se somehow plays a role in collapsing the wavefunction, although this was a possibility that did intrigue a few prominent physicists and continues to be echoed in New Age metaphysics and in some corners of parapsychology. Most physicists now agree that observation can include just an ordinary interaction with the environment; it's just that experimental measurements (observations) happen to be the only interactions with the environment that the physicist is able to measure and record. Thus, when physicists talk in terms of observation or measurement, they are being precise and cautious, even if the famous "measurement problem" might refer to what happens in *any* causal interaction in the natural world.

Thanks mainly to Bohr's forceful personality and his stature in the field, physicists for decades, especially through the height of the Cold War when their services were funded for defense, were told to not question, not to hunt for other explanations or other interpretations for (what looked like) randomness in the quantum realm. But there have always been dissenters, and dissent has grown in the past few decades.

The cultural meme that best expresses the quantum no-man's-land prior to observation, according to the Copenhagen interpretation, comes from a single, enormously famous paragraph in a 1935 paper by Erwin Schrödinger: A cat is put in a box with a vial of poison and hammer connected to a Geiger counter and an atom of an unstable radioactive isotope. If the atom decays (as detected by the Geiger counter), the hammer breaks the glass and kills the cat; if it doesn't, the cat lives. Until someone opens the box to see what happened, both possibilities exist in a state of what is called *superposition*. It's not simply that the cat is both alive and dead; the cat's fate, said to be entangled with the radioactive isotope, is indeterminate until observation collapses the wavefunction that describes their inseparable fates.*

All this business of alive/not-alive cats makes for funny jokes and tee-shirts, but there's really nothing all that funny about it. In 1961, one of the aforementioned dissenters, a Hungarian physicist named Eugene Wigner, extrapolated on Schrödinger's thought experiment to show the flaws of the mainstream Bohrian view: Your friend, the owner of the possibly unfortunate cat, goes into the room with the box to check on her animal, but shuts the door behind her. While you wait outside, with no indication of what is happening in the room, the superposition of the dead/alive cat remains, but the same superposition now describes your sad/relieved friend.

The quantum dogma of the time dictated that your friend, or at least her emotional state, are now part of the same wavefunction as the cat—they are entangled now, described by the same equation … at least, for you. Inside the room, however, your friend has probably already torn open the box and collapsed the wavefunction describing the cat and the radioactive atom *for her*. This creates a

---

* See Karen Barad, *Meeting the Universe Halfway* (Durham, NC: Duke University Press, 2007).

contradiction, and thus it is an indication that something is fundamentally wrong with the Copenhagen interpretation.* Incidentally, Wigner was among the minority partisans of the idea that it is consciousness itself that collapses the wavefunction.

Another dissenter was Einstein, who had always disliked Bohr's insistence on some probabilistic essence of quantum reality, insisting that "God does not play dice." He was not a believer in real wavefunctions or the necessity of observers, let alone conscious ones. He thought that there had to be a definite real world out there, preceding and independent of our observations, and that there was just something we were missing. It was to show the problems with the mainstream view that Einstein and two of his colleagues, Nathan Rosen and Boris Podolsky (collectively, "EPR"), proposed that idea of *quantum entanglement*: Two particles can be specially prepared in such a way that even if you send them far away from each other, measuring one particle causes the other particle to have a same or corresponding measurement—even if they are light years apart. The most intuitive way of understanding this was that the particles somehow communicated, as if by instantaneous telepathy, thus seeming to violate light's finite speed limit according to Einstein's special theory of relativity.

---

* As you can imagine, you can create an infinite number of nested rooms this way, infinitely expanding that wavefunction by adding friends and friends of friends, perhaps even some enemies, and end up with a completely solipsistic ontology. See Amanda Gefter, *Trespassing on Einstein's Lawn* (New York: Bantam Books, 2014).

The EPR paper did not have the intended or expected effect. Rather than force a revision of mainstream quantum physics, the field adapted to the new reality of quantum-entangled particles, and experiment after experiment confirmed that entanglement was real, whatever it meant. But decades of physics research in the 1980s and 1990s—including research by Nick Herbert—established that the intuitive "information transmission" view of what happens in an entangled system is probably not right; entanglement cannot be used to send a message.* The nonlocality idea—another that has been embraced widely in popular metaphysics and has been invoked by parapsychologists to explain reports of telepathy—is by no means the only, or most persuasive, interpretation.

Was the answer lying in older, contrarian views that had been overshadowed by Bohr's towering personality? Already in the 1950s, a French physicist named Olivier Costa de Beauregard had proposed an alternative explanation for entanglement: Instead of a telepathic-like connection, it is, he suggested, a function of a kind of causal zig-zag back and forth through time.† Interacting with one of a pair of entangled electrons "sends" information back in time to when they became entangled, and thus that information becomes part of the backstory of its twin. Recall that something like this had already been proposed by Walter Schottky in the early 1920s, and in fact

---

* David Kaiser, *How the Hippies Saved Physics* (New York: W.W. Norton & Company, 2011).

† Huw Price, Ken Wharton, "A Live Alternative to Quantum Spooks," arXiv:1510.06712v2, November 5, 2015.

another physicist named Satosi Watanabe also proposed a similar time-symmetric interpretation of the quantum measurement problem in 1955, but … again, crickets. Costa de Beauregard's "Parisian zig-zag," too, had to wait until after the Cold War to get dusted off and reconsidered.

Murray Gell-Mann, in a 1976 Nobel lecture, declared that Bohr "brainwashed an entire generation of physicists" with his Copenhagen interpretation.[*] And more and more physicists in the last few decades have expressed similar sentiments. In a recent history of Bohr and his influence, David Ferry calls the Dane's century-long domination of physics the "Copenhagen Conspiracy."[†] And David Griffiths wrote in the Afterword to his widely taught *Introduction to Quantum Mechanics* that "It is entirely possible that future generations will look back, from the vantage point of a more sophisticated theory, and wonder how we could have been so gullible."[‡] That more sophisticated theory may well be one of the retrocausal interpretations that are lately gaining ground in physics. There is mounting evidence that Schottky and Watanabe and Costa de Beauregard might have been right, that events in the quantum world are co-determined by both past and future events or interactions.

---

[*] Quoted in Larry Hardesty, "Fluid Mechanics Suggests Alternative to Quantum Orthodoxy," *MIT News*, September 12, 2014, https://news.mit.edu/2014/fluid-systems-quantum-mechanics-0912;

[†] David Ferry, *The Copenhagen Conspiracy* (Singapore: Pan Stanford Publishing Pte. Ltd., 2019).

[‡] In Hardesty, "Fluid Mechanics."

Retrocausation means that in addition to the efficient causes we intuitively understand and perceive that determine an object in the rightward direction in the block universe, from past to future—like balls hitting each other in sequence on a billiards table—objects' behavior may also be affected, inflected, or just constrained by influences ahead in time. It would mean that the "uncertain" component of the behavior of a photon or electron actually reflects some influence exerted by the next particle it will interact with, propagating backward across its worldline. Such backward or pastward influence may tend to be negligible and unnoticeable on our (classical) scale, where events mostly unfold according to entropy. The one-way "arrow of time" that we naturally experience has been explained as our natural attunement to entropy or to entropic processes in the brain. But in the tiny quantum world, this backward influence may be considerable, and there are certain special circumstances when it might "scale up" and be relevant in the classical world too.

Retrocausation or some kind of future determination (there are different ways of talking about this) has a growing number of advocates in the physics community. Israeli physicist Yakir Aharonov resurrected Watanabe's time-symmetric interpretation, renamed it the "two state vector formalism," and has collaborated with Chapman University physicist Jeff Tollaksen on a number of recent papers reinterpreting physics staples like the "double-slit experiment" in time-symmetric terms. (More on that experiment later.) In the 1990s, Cambridge University philosopher Huw Price wrote a brilliant book, *Time's Arrow and Archimedes' Point*, arguing for what he called "advanced action" (retrocausation) as an explanation for entanglement, Costa de Beauregard's Parisian zig-zag.* He also made the interesting observation that it is philosophical matters like the free-will problem that have been major barriers to acceptance of this idea—strange, if you buy into the assumption that scientists are led by, you know, science, and not their personal beliefs or biases. More recently, Price and San Jose State University physicist Ken Wharton have coauthored several articles in journals and more popular venues elaborating an advanced-action interpretation of entanglement.†

Somewhat similar to Price and Wharton, but using a different vocabulary, an interdisciplinary team at Elizabethtown College, Michael Silberstein, W. M. Stuckey, and Timothy McDevitt, have argued for a framework they call "relational block world." Taking a God's-eye position on space, time, and matter, they argue, forces us to see the future as just as determinative as the past, and they compare their framework to the non-time-bound worldview of the alien heptapods in Denis Villeneuve's 2016 movie *Arrival* or the Tralfamadorians in Kurt Vonnegut's 1969 novel

---

* Huw Price, *Time's Arrow and Archimedes' Point* (Oxford, UK: Oxford University Press, 1996).

† Nobel laureate Roger Penrose now talks explicitly in terms of retrocausation too. See Andréa Morris, "Testing A Time-Jumping, Multiverse-Killing, Consciousness-Spawning Theory of Reality," *Forbes*, Oct 23, 2023, https://www.forbes.com/sites/andreamorris/2023/10/23/testing-a-time-jumping-multiverse-killing-consciousness-spawning-theory-of-reality/.

*Slaughterhouse Five.** For simplicity's sake here, I consider their model retrocausal, even though in their book *Beyond the Dynamical Universe* they studiously avoid dynamical-sounding verbs like "cause" in favor of a more static language of future "constraint."

Whatever exact terms are used, these frameworks force us to accept the mind-bending idea that the future affects or co-determines the present. But they are elegant and parsimonious, and they greatly demystify the famously "spooky" quantum world. They restore locality rather than allowing a weird and inexplicable exception to relativity, and this could help pave the way for the long-sought theory of quantum gravity that would reconcile the worlds of the big and the very small. These frameworks also turn Schrödinger's cat into a story about our own present ignorance of something definite that happened when we were not looking, not a real ghostly neither/nor state. But they also confront us with a truly mind-blowing explanation for why we couldn't fully predict what would happen in that box: Somehow, it was our later observation that would turn out to be the missing piece of the causal puzzle, exerting a *retro-influence* on the events inside. Our peering into the box was part of what *had already* decided the cat's fate.†

---

* Michael Silberstein, W. M. Stuckey, Timothy McDevitt, *Beyond the Dynamical Universe* (Oxford, UK: Oxford University Press, 2018).

† Ken Wharton argues that one natural phenomenon that makes little sense unless we accept some kind of retrocausation is a curious thing called *frustrated spontaneous emission*: a radioactive substance will not decay if it is placed in an environment where its emitted radiation cannot

Observation, in this new view, is not reality's midwife. Rather, reality was always definite, always adult. But part of that definiteness—the part that made it unpredictable—was provided by that observation. This retro-determination of things by observation is truly a "third way" that carries us beyond the old opposition between Einstein and Bohr, between "observation discovers something pre-existing" (Einstein) and "observation creates something out of nothing" (Bohr). Instead, observation supplies what was missing (uncertain) about the particle *before* it was measured. But again, observations are causal interactions we happen to know about; really, all interactions in the material world may do the same thing: constrain or influence the prior behavior of the objects interacting.

Predictors

I tend to look at the history of science through a Freudian psychoanalytic lens, and to me, the Copenhagen interpretation has all the hallmarks of a collective hysterical defense mechanism against the reality of things going backwards. I really suspect that, in the future, once we are more fully acclimated to the retrocausal, time-looping reality of things, we may indeed look back on Bohr's dogma and shake our heads at the absurdities (cats that are alive and dead at the same time, particles communicating

---

be absorbed—suggesting it somehow "knows" there is something there to receive it. See George Musser, "The Quantum Mechanics of Fate," *Nautilus*, January 30, 2014, http://nautil.us/issue/9/time/the-quantum-mechanics-of-fate.

telepathically, and so on) that generations of physicists desperate to keep things linear were prepared to accept and that a quantum-curious public took on faith.

But one of the articles of faith that physicists—even the retrocausalists—*still* assert, to this day, is that it ultimately doesn't matter which interpretation of the quantum measurement problem is adopted. It is just a matter of taste (allegedly) whether we imagine that causes can go backwards, things are just random, we really live in a quantum multiverse, or some other picture.[*] The math still works out the same, the equations for predicting what will happen in quantum systems still give the same answers. But I suggest that this is another part of the defense mechanism: If causes really can go backwards, and if this can be leveraged in the world of technology or has already been leveraged in, say, biology, then the implications are *huge*. It is a paradigm that could shift the world, and ourselves.

Testing retrocausation experimentally is extremely difficult, though—another reason, besides taboos, that this framework has languished for so long.[†] Except in rare circumstances, we can't know exactly what those future causes or influences are, because it's not the future yet—which is exactly why they have seemed random (or

---

[*] John Gribbin's book *Six Impossible Things* helpfully summarizes the most mainstream opinions on this question, including one quasi-retrocausal theory, John Cramer's "transactional interpretation." See John Gribbin, *Six Impossible Things* (Cambridge, MA: MIT Press, 2019).

[†] Price, *Time's Arrow*.

causeless) to our eyes, stuck as we are in the present. But when you put constraints on those future causes—a technique now being used widely in experimental research called "post-selection"—and when you combine this with a new experimental method called "weak measurement" that uses ultrasensitive measurement tools, it now seems to be possible to investigate retrocausation.

For instance, in 2009, at the University of Rochester, a research team led by John Howell passed a laser through a beam-splitter to create two identical beams. The amplitudes of both beams were weakly measured using a special type of interferometer, a mirror with an extremely sensitive mirror that could detect the subtle pressure of the photons bouncing off it. Then, only one of the two beams was post-selected to receive a later "strong" measurement. The results of the initial weak measurements suggested that the photons in the one later-measured beam were considerably amplified compared to the beams in what, if this were a psychological or biomedical experiment, would be called the control group. Although the researchers didn't make explicitly retrocausal claims for their finding, University of Bristol quantum physicist Sandu Popescu, commenting on their paper, was less cautious: The experiment seemed to demonstrate the two-way flow of time and causation. The later measurement somehow amplified those photons *in their past.**

---

* P. Ben Dixon, et al., "Ultrasensitive Beam Deflection Measurement via Interferometric Weak Value Amplification," *Physical Review Letters* 102 (2009): 173601-1-4; Sandu Popescu, "Weak Measurements Just Got

Other similar experiments seem to show the same thing. In 2016, two researchers including a physicist at the Naval Surface Warfare Center in Dhalgren, Virginia published results of an experiment in which they weakly measured photons in a split laser beam using a different kind of ultrasensitive interferometer, again finding changes to the photons in one of the beams in an initial weak measurement that could best be explained by a subsequent interaction in the experimental setup. They interpreted their findings retrocausally, in light of Aharonov's time-symmetric ("two state vector") formalism.*

In 2011, MIT engineering professor Seth Lloyd and several colleagues used an experimental setup involving post-selection and what they called a "quantum gun" to shoot a photon at itself in its past. Although it was only a few billionths of a second into the past, the results seemed to support that particles like photons can be influenced backwards in time by processes occurring subsequent to some outcome of interest. And of course, history remained self-consistent, proving that grandfathers (and past selves) "are tough guys to kill."† As reported recently in *New*

Stronger," *Physics* 2(32),April 27, 2009, https://physics.aps.org/articles/v2/32.

* S. E. Spence, Allen Parks, "Experimental Evidence for Retro-Causation in Quantum Mechanics Using Weak Values," *Quantum Studies: Mathematics and Foundations* 4, July 22, 2016, DOI:10.1007/s40509-016-0082-x.

† Seth Lloyd et al., "Closed Timelike Curves via Postselection: Theory and Experimental Test of Consistency" (*Physical Review Letters* 106:040403), 2011.

Scientist, Lloyd is now planning a further experiment that elaborates on these ideas.*

In 2021, Julia Mossbridge reported "causally ambiguous behavior" in trials she had conducted using a different kind of optical system that just shoots photons at a detector, through slits in a screen. It is the kind of setup used for variants of the famous double-slit experiment that demonstrate light's wavelike properties by making an interference pattern on the detector. There are especially intriguing versions of this experiment that show that photons seem to interfere with each other even when they are shot through the slits at different times, one by one, and this is the kind of setup Mossbridge used. What she found was that the interference pattern produced was different depending on how long she ran each trial of her experiment—that is, how long the apparatus was turned on. She couldn't say conclusively that what she was detecting was photons entangled with each other across time, but that is one possible interpretation of the results she got.†

And as I was writing this book, a team at the University of Toronto reported the results of an incredibly intriguing discovery that prior research had already hinted

---

* Miriam Frankel, "Quantum Time Travel: The Experiment to 'Send a Particle into the Past'," *New Scientist*, May 29, 2024, https://www.newscientist.com/article/mg26234932-900-quantum-time-travel-the-experiment-to-send-a-particle-into-the-past/.

† Julia Mossbridge, "Long Time-Frame Causally Ambiguous Behavior Demonstrated in an Optical System," ResearchGate, February, 2021, https://www.researchgate.net/publication/349106030_Long_time-frame_causally_ambiguous_behavior_demonstrated_in_an_optical_system.

at: Photons entering a cloud of supercooled rubidium atoms spent a "negative amount of time" in that medium before exiting. It means that they seem to have left the rubidium cloud before they entered.* If your mind is not blown by *that*, it should be: Go take a moment, make yourself some nice calming tea if you'd like, and reflect.

These experiments all remind me of a very short story by Ted Chiang, one of those rare science-fiction writers who address the question of time travel and its variants (e.g., precognition) rigorously and realistically. "What's Expected of Us" (in Chiang's collection *Exhalation*) is about a simple fidget-spinner-like toy called a Predictor, with a button and a light linked by a negative time delay—meaning it lights up a second *before* the button is pressed. With no way to trick or cheat the device, some users are troubled (to put it mildly) by what it seems to reveal about the nonexistence of free will. But the retrocausal behavior hinted at by the above experimental findings suggests that such a device could become a reality.

Consider how useful it would be to harness some "negative time delay" in safety devices: for instance, a quantum airbag in your car that deploys a second in advance of a collision that has not been anticipated by the vehicle's standard safety systems, or an ejector seat that throws a fighter pilot to safety seconds before their plane is

---

* Daniela Angulo et al., "Experimental Evidence that a Photon Can Spend a Negative Amount of Time in an Atom Cloud," arXiv:2409.03680v1 [quant-ph] September 5, 2024, https://arxiv.org/pdf/2409.03680.

struck by an anti-aircraft laser. Predictive circuits in medical implants could sound an alert or deliver stabilizing medication immediately prior to an imminent cardiac arrest.

Note that I'm not talking about ordinary prediction, which gathers available evidence to calculate a probable or possible outcome, but a device that pre-sponds to an "already existing" future event. For the same reason you cannot change the timeline, you could not use such a device to prevent the pre-detected outcome. If the outcome never happens in the future, then there is nothing to foresee—it's a permutation of the grandfather paradox. But you can prepare for an outcome to either mitigate its impact or be ready to capitalize on it when it occurs. As one might imagine, the first use of such devices will be to predict the outcomes of sporting events and fluctuations in the value of shares on a stock market. The first entrepreneur or company to build a reliable "future detector" will get very rich, very quickly, at least before the authorities detect something fishy afoot.

Possibilities, Singularities

So, let's return to that idea Gibson envisioned in *The Peripheral*: a device that allows communication backwards as well as forwards through time. We already live in a wired world, but soon, the wiring that connects people across space—through emails, phone calls, virtual reality, and telepresence—may really be able to connect us across the fourth dimension, too. As Gibson envisioned, quantum computers could be the key.

A quantum computer is a matrix of particles like electrons that are entangled with each other, effectively scaling up the mysterious properties of the quantum world and exploiting them for information-processing purposes. Each of those particles is analogous to a "bit" in a classical computer; but instead of just having two possible states, 0 or 1 (off or on), each "qubit" as it is called (for quantum bit) can be in that Schrödinger's-cat state of superposition, many different states or no-states at the same time. It is said to enable a sort of hyper-parallel processing where the same cubits perform multiple computations simultaneously.

A quantum computer with just a small number of qubits can perform calculations that would take an ordinary computer years or centuries or longer. This super-computational ability has even given rise to the belief among some theorists, most notably Oxford physicist David Deutsch, that these devices actually tap into other, parallel universes. This is the "many-worlds" interpretation that was formulated in 1957 by a Princeton physics graduate student named Hugh Everett III. It is the serious, physics counterpart of the fictional parallel worlds of Murray Leinster or Philip K. Dick. Serious, certainly, but I argue (and more and more physicists would agree), not necessarily credible.

Like some of his elders, Everett had been bothered by the inherent randomness of fundamental reality insisted upon by the then-mainstream Copenhagen interpretation. Everett had studied under Eugene Wigner at Princeton while getting his masters, and then worked with John Wheeler while doing his doctoral research. (We'll hear much more about Wheeler in Chapter 5). It was in his dissertation that Everett offered his "relative-state formulation," which came to be renamed "many worlds" only later. He proposed that at every decision point in nature, whenever a particle is interacted with or measured, there could be practically infinite other universes where it took a different direction or interacted in a different way. The universe, in this theory, is really a multiverse, a garden of effectively infinite forking paths. Our world is just one of those ever-branching-off paths.

Like those time-symmetric interpretations I mentioned earlier, Everett's idea was ignored for more than two decades. But it grew in popularity in the 1990s with the work of Deutsch and others, who found it to be an intuitive way of understanding the fact that a quantum computer can quickly solve problems that seem to greatly exceed its finite computing power—its number of qubits—at any given point in time. It also has other charms: It makes quantum causation feel an iota less senseless, less random, without requiring we go the full distance into weirdness by adopting a retrocausal view. And, unlike retrocausal models, it satisfies our need to feel like our will is free and that, even if there is a block-like future ahead of us, the world is full of unrealized, unactualized possibilities, paths not taken—even if those paths not taken are sealed off in other blocks. And many worlds has also been invoked by some physicists who remain unconvinced by Igor

Novikov's self-consistency principle, as a way to allow time travel without causing paradox.*

However, after its brief heyday (probably abetted by numerous movies on the multiverse theme), we may be seeing the waning of the many-worlds interpretation. Physicists are abandoning it, more and more.† Among other things, there's no actual evidence for it, despite the breathless claims of entrepreneurs and venture capitalists in the quantum computing field.‡ And it conflicts with hidden-variables theories like retrocausation. If the quantum world admits causal influence from the future, then the randomness that troubled Everett enough to propose a branching-fork view of quantum measurement goes out the

---

* See Barak Shoshany, Jared Wogan, "Wormhole Time Machines and Multiple Histories," arXiv:2110.02448v3 [gr-qc], March 15, 2023. Shoshany and Wogan base their argument on various grandfather-type paradoxes that can really be dismissed as situations that can't happen simply because they didn't. They are also overly bothered by bootstrap scenarios, mistaking tautologies for paradoxes.

† See Michael Brooks, "How a Quantum Innovation May Quash the Idea of the Multiverse," *New Scientist*, January 6, 2025, https://www.newscientist.com/article/mg26435252-200-how-a-quantum-innovation-may-quash-the-idea-of-the-multiverse/; Sabine Hossenfelder, "Does the Many Worlds Interpretation make sense?," YouTube, October 21, 2023, https://www.youtube.com/watch?v=433tAfO4dbA.

‡ As I was finishing this book, Google's Quantum AI team lead Hartmut Neven invoked many worlds to account for the success of a new chip called Willow that, it was claimed, had solved a problem in 5 minutes that would have taken longer than the age of the universe for an ordinary supercomputer to solve. Hartmut Neven, "Meet Willow, our state-of-the-art quantum chip," *The Keyword*, December 9, 2024, https://blog.google/technology/research/google-willow-quantum-chip/.

window; and if the entanglement that is intrinsic to a quantum computer is really about retrocausation, then so does the need for those other, parallel universes.

Despite its superficial appeal, I personally think many worlds is a bit of a cop-out, as well as a symptom of our human preference to think in terms of space (or in this case, spaces) than in terms of time. If there's insufficient volume in your room right now for something like a computation, it is easier for us to imagine it spilling over into other, unseen rooms than to imagine that what is happening in your room *in the future* might be relevant. Sideways in time is simply easier for us stubbornly 3D humans to picture in our heads than both forwards and backwards across a dimension we cannot see.

Per the Parisian zig-zag of Costa de Beauregard and Price and Wharton, those entangled qubits are entangled with each other and themselves across their histories, in which case the future and past could well be where all that extra (missing) computing power is to be found. Quantum computers may fast-forward (or "tunnel" in the idiom of quantum physics) to the right answer, thanks to their ability to compute across large swathes of their history, or their entire history.

There would be ways to actually test this hypothesis: The computational power of a quantum computer would depend on how long you leave it plugged in—a little like Mossbridge's tabletop experiment with entangled photons. If this hypothesis is right, duration of activation would make a difference, and comparing two otherwise identical quantum computers that are left activated for different lengths of time would show this difference. The longer-lived computer would be more powerful.*

Here's the main thing: If quantum computers are really reaching into their future for the answers to problems, then it transforms these devices into something much, much more than amped-up number crunchers, and something much more interesting. Multiple research teams have found that causal order can be reversed using entangled qubits.† A simplistic way of putting it is that you could theoretically have a kind of output that preceded an input. In which case, such a device might be used not only to scale up the retro-signaling capabilities of something like a Predictor by many orders of magnitude, but could also serve as a kind of interface, a chronophone or "c-mail" server, connecting future and past users—more or less as Gibson envisioned in *The Peripheral* but minus the branching-fork continua.

Think of the possibilities: Weather satellites connected to a quantum server could alert meteorologists *in the past* about destructive hurricanes, which would give affected communities time to prepare or evacuate. Seismologic and oceanographic sensors could do the same to help mitigate the impact of earthquakes and tsunamis. And of course, if

---

* In *Time Loops*, I speculated that this could be true of biological systems as well: Longevity might be causal in the oft-found association between longevity and intelligence.

† Jayne Thompson, et al., "Causal Asymmetry in a Quantum World," *Physical Review X*, 8 (2018), 031013, DOI: https://doi.org/10.1103/PhysRevX.8.031013.

the police or national guard could get forewarnings of terror attacks, those too would become less threatening.

Once again, by sending information back in time, you are not able to *change* history and prevent the threat before it occurred—a conceptual bug implanted by all those science-fiction novels and films of your youth. A quantum-computer-mediated version of the Precrime Unit from Philip K. Dick's story "The Minority Report" would not, for instance, be able to actually prevent a foreseen crime from occurring. But you would be able to mount a preparatory response to reduce or eliminate the damage or harm, or be ready to capture the criminal afterward. These would be small examples of what I called history gardening: Using time-travel technology to create conditions in the past for a better future that would have been unattainable otherwise.

The possibilities only multiply if you leave such a device plugged in for years or decades. For instance, picture the following scenario: You are an epidemiologist in the late 21st century working for NICOID, the National Institute on Coronaviruses and Other Infectious Diseases. Our fragile world, reeling from the effects of runaway climate change, is now a tropical hothouse of rapidly evolving pathogens. There is no time to meet each new pandemic threat—new lethal viruses and antibiotic-resistant bacterial strains are identified practically every month. But unlike earlier historical pandemics that you learned about in your school history holograms, the new ones are less of a danger thanks to that same kind of quantum server that saves people from hurricanes and earthquakes. Data on newly identified pathogens is fed to laboratories in the past, giving them time to develop and test vaccines and antidotes before they spread widely. Each new virus and superbug finds that humans' immune systems are already pre-prepared, and the diseases fizzle before they are able to gain a foothold.

These are somewhat simplistic scenarios mainly to spark your imagination—and again, our imagination doesn't have to be exactly accurate; it just needs to motivate us and prepare us for possibilities. One problem with these scenarios is that they presuppose benevolent governmental authorities that place the safety and welfare of their citizens topmost. The reality is that if and when these technologies are developed, they will be (or, who knows, perhaps already are) the most secret of the most secret for the nation states and other actors possessing them, and strategic and security priorities will in all likelihood trump many of the possible social benefits, at least initially.

Quantum servers could transform warfare, for one thing, with rival factions tweaking the past to produce favorable outcomes for their side in the immediate or near future. In geopolitical conflicts, advance knowledge of many enemy actions is not acted upon—and human pawns are sometimes sacrificed—in order to keep the intelligence source secret. Already there is a wide suspicion among UFO researchers that transformative time-travel-related technologies like antigravity have been developed but are either hoarded for secretive military purposes or suppressed by entrenched industries fearful of their own obsolescence. State users will be very loathe to tip their hand. When you have rival countries and factions vying for supremacy and gaming the time-travel system, things could get very complicated. They might already be complicated.

Also, there are obviously many other nefarious or at least amoral possible uses of such technology. In *The Peripheral*, for instance, past histories in alternate continua are exploited as test beds for unethical and lucrative biomedical experimentation. The people in those alternate histories are expendable lab rats, and calling their worlds "stubs" psychologically makes it easier to "third-world" them.* But that kind of trans-temporal exploitation is perfectly possible within a single historical timeline. For instance, an amoral pharmaceutical company could test new drugs on some unwitting population in its past. Defense contractors could test new weapons the same way. History shows that greedy countries and corporations hardly need other timelines; they can and do "third world" the third world, all the time. The past—and remember, we here now are "the past" of our timefaring future—could readily be a new kind of third world.

Whether the kinds of quantum servers I've described are still on the horizon or already exist in some form, deep within the black world, in the hands of foreign governments, or in the island lairs of Bond-villain-style oligarchs, the first ones that are switched on and protected from disconnection could be used not only to gain military and financial advantage over adversaries but also to borrow

---

* Gibson credits the idea of exploiting the resources of alternate histories to a 1985 short story by Bruce Sterling and Lewis Shiner, "Mozart in Mirrorshades."

further technological knowhow from future users. This could make runaway technological development envisioned and feared around AI seem almost trivial, and it is another reason why holders of this technology may try hard to keep it secret as long as they can. In fact, it really wouldn't be technological "development" in the usual sense but spontaneous emergence of new knowledge arising from a kind of informational bootstrapping—the same kind of "Jinn" depicted by James Cameron in his *Terminator* films, where debris from future AI discovered by a defense contractor is used to back-engineer that very AI, in a loop.

Again, such causal/informational bootstrapping is perfectly allowed—and might even be pervasive—in a universe that includes retrocausation. There's nothing paradoxical about things and ideas arising from nothing—it just really bugs us. The "things" part is well known in physics, in fact: There's the Big Bang itself, first of all, which was the biggest things-arising-from-nothing in history. (We'll come back to that, and how it really may have been part of a time loop, later.) But even right now, as we speak, the vacuum of empty space is a seething "quantum foam" where particle-antiparticle pairs arise out of nothing and then crash together with a great release of energy. They might not really be "pairs," though: One hypothesis is that it's really a single particle looping forward and then backward in time.

Such loops might underlie the creation of new ideas. My previous books have explored the implications of the brain as a kind of semi-quantum computer constantly reaching into its future for information. Artists and scientists and inventors get ideas from their future selves in dreams or in creative flow states, helping explain the ubiquitous stories of prophetic books and artworks predicting future events and experiences. In their work, they then contribute to bringing about the future situation that inspired them. The implication of this is that there is no point in an artist's life where some new idea is "hammered together" out of already-available ingredients on some inner cognitive workbench. Original ideas are quite literally Jinn, bootstrapping themselves into existence from loops in a creator's life.*

(I see that skeptical side-eye you're giving me. You should celebrate: It means you, as a thinker and creator, are a bit like a god. Or a lot.)

With scientific and technological innovations, we could think of it, again, almost as a kind of fast-forwarding or tunneling, skipping over the intermediate steps that our present, linear understanding of science and knowledge rests on.

Given that a future user of such technology cannot send a message to an earlier point in time than when the computer was switched on, it creates something like an event horizon around future technology. Michael P. Masters is a physical anthropologist at Montana Tech and the leading exponent of the time-travel interpretation of the UFO phenomenon, and we'll delve into his work in the next chapter. But in interviews, he describes the devel-

---

* Wargo, *From Nowhere*; Eric Wargo, *Where Was It Before the Dream?* (forthcoming).

opment of physical time travel as potentially creating a "super-technological singularity": Future-human time travelers visiting us in the present could be delivering future technology, or at least, leaving it behind, enabling us to reverse-engineer it.* Many of the questions that came to the fore among UFO researchers due to congressional hearings and whistleblowers in 2023 and 2024—rumors of flying saucer crash retrievals and back-engineering programs—spark imaginings of such *Terminator*-like possibilities. But informational time travel via quantum servers or perhaps other, simpler, future-detection technologies may be all that is needed to precipitate such a singularity.

If our civilization is indeed approaching such a horizon, it could mean the end of ordinary science as we know it. The linear evolutionary processes through which, Karl Popper insisted, scientific knowledge accumulates, may be replaced by a new informational regime in which the secrets of nature are just given to us by our descendants or our future selves—not in dreams but through the kinds of technologies I mentioned.

When Arthur C. Clarke wrote that any sufficiently advanced technology will be indistinguishable from magic, he may have been more right than he knew. What if sufficiently advanced technology may literally *conjure itself into being* in the no-space and no-time of time loops, without anybody doing a lick of intellectual or technical

---

* "Evolution Expert: Aliens Are Future Humans & UFOs are Time Machines—Mike Masters," Danny Jones podcast, YouTube, Sep 16, 2024, https://www.youtube.com/watch?v=wrucxSNvYQY.

work as we usually understand it? Work is a thermodynamic thing, but our timefaring future may literally be a post-work world, with new technological boons, like Jinn, popping into being from nowhere and from nothing.

The fact that these transformations will be (/already are?) unevenly distributed and possibly kept extremely secret by those who hold them is one reason to pay special attention to the simpler, Predictor-type future-detecting devices I mentioned. These could be more accessible to us little people who live and work far outside the deep-black world of hypersecretive aerospace companies and government agencies with their billion-dollar, supercooled quantum computers. With the help of some experiment.com donations, Mossbridge was able to purchase the instrument for her time-traveling photons experiment from a science equipment company. Once our eyes are opened to the possibilities, there may be myriad other mechanisms for future-detectors hiding in plain sight, including even in nature. Again, your brain is probably such a device; but so, probably, are much simpler biological systems. The latter may be the simplest to work with: Tie some anticipatory behavior of a group of mice or plants or algae to an outcome of real-world interest, and presto, you have a future detector. All it may take is imagination, some biological know-how, and the willingness to set aside the cultural stigma of teleology.

The (Post-)Human Factor

Besides the temporal horizon I mentioned (not being able to send a message to anytime earlier than the server was plugged in and activated), there are other important limitations of purely informational interaction or communication across time. These hinge on decidedly non-technological factors of the humans or posthumans doing the communicating.

One is the issue of trust: How will recipients of a message in the present trust that the future sender (even their future self) is who they say they are, or that the sender has their best interests at heart? How can we trust that the server will not have been hacked at some point farther ahead and that, say, a forewarning about some impending opportunity or disaster is not misdirection designed to lead us astray—perhaps a disinformation gambit in some time war? Mossbridge discusses the complexities of trust in time travel—and how to build trust across time through unconditional love—in a fascinating article, "How to Avoid the Time Wars."*

The issues go beyond trust, though, and they carry us back to those paradoxes that don't exist but haunt our thinking anyway: Information cannot reflux back in time in a way that would lead to its foreclosure in the future. Often, though, we want it to—and that's even true, oddly enough, when outcomes are positive. Our neurotic assertion of our own free will demands we always try to change (past or future) history or otherwise "do something new." This is why I always point out that in understanding retrocausation, time travel, and time loops, Freud is just as important as Einstein. Not unlike the way planets and

---

* Mossbridge, "How to Avoid the Time Wars."

asteroids and comets take the easiest (straight) path through curved spacetime, messages that are successfully sent across time take paths of least resistance through a curved landscape of sense and sensibility. It means a past user of a quantum computer might not be able to get a clear, unambiguous, completely trusted message from their future self.

A beautiful cinematic dramatization of this principle is Christopher Nolan's 2014 film *Interstellar*. Near the end, astronaut Coop (Matthew McConaughey) is able to communicate with his younger self (as well as his daughter) decades in the past using a tesseract created by our far-future descendants. Coop's messages are oblique—he can't address his younger self directly but has to send symbolic messages using the books on his bookshelf. Early in the film, his younger self, before his planned mission to a black hole, realized there was something significant occurring with those books but missed the precise meaning: a warning not to go. (I argue elsewhere that this logic perfectly encapsulates that of precognitive dreams as well: They are messages from our future that are oblique and symbolic and almost inevitably misunderstood until after the experience precognized.*)

An additional difficulty might occur to you: A user in the future might not be sufficiently motivated to send a message into the past to initiate an action there/then that they believed had already been done and accomplished in

---

* Eric Wargo, *Precognitive Dreamwork and the Long Self* (Rochester, VT: Inner Traditions, 2021).

their present. They might instead lazily "bilk" the fore-ordained action … which of course, would prevent it from ever having been accomplished in the past. Thus, this is another kind of situation that simply would not occur, because it didn't already. But considering our reaction to such a thought experiment can be revealing, as it helps us understand some of the deep-seated anxieties and resistances to time travel.

We seem to be hardwired, perhaps as a legacy of millions of years foraging and hunting for food, to be motivated by the unexpected and unknown. Our brains' reward system thrives on moments when our expectation is undercut somehow, and dopamine drives us toward the incentive of discovery. Building on this rewarding signal of the unexpected is the further payoff of seizing new opportunities and tinkering with the expected. It is probably our natural revulsion to being forced to do something that someone else (including our own future self) has mandated that leads to the natural thought process of wanting to thwart (or bilk) history, the specific madness of Stephen Hawking's mad scientist and, really, the madness underlying all of those misleading thought experiments about changing the past. People need to feel free, to not feel constrained or like they are going through motions. Feeling like you're acting a part in a play is a form of dissociation that is symptomatic of depression.

This is the issue that Chiang tackles in "What's Expected of Us." A third of Predictor users slump into a kind of silent immobility called *akinetic mutism*, as playing with this time toy has destroyed their prior belief in their own free will. People find the block universe where the future already exists fascinating, but at the same time, they are troubled because it seems to take away that sense of uncertainty that gives life purpose and motivates us. But again, in the block universe of Einstein and Minkowski, your freely-willed-feeling actions right now are already in the past from the viewpoint of another observer. It leads to the thought of those akinetic mutists: Why get up in the morning and do anything? Why lift a finger, since everything will happen anyway?

A moment's thought reveals the fallacy, the bad faith, of akinetic mutism, though: If you don't reach and pick up your spoon, your oatmeal will not magically levitate to your mouth. The predestined future includes all your "freely willed" actions that bring it about. There is no escaping the responsibility of action. Harboring the idea that you should "just lie in bed because what's the point?" is to be beguiled by an *image*, a stage-representation of passivity, not anything real.

Not being driven to a state of depression (or madness) by the self-consistency laws governing the universe will be part of what is culturally and perhaps even genetically selected for in our timefaring future. Partly, though, timefarers may arrange things to avoid putting themselves and each other in situations where they are called upon to do things that they know have already been accomplished. This includes weird scenarios in which they may encounter their younger or older selves—a topic we'll return to in the final chapter.

For the same reason, to work reliably, trans-temporal communication may need to be arranged with the kinds of double-and triple-blinding protocols that you would find in psychology experiments. The sender of a future message would be unknown to the past recipient, and vice versa, and knowledge of already-accomplished outcomes might be kept secret from agents tasked with seeding them in the past. In fact, in many ways it will be advantageous to simply automate things, taking the neurotic humans out of the system altogether … but that of course could be dangerous for all the reasons Cameron envisioned with his *Terminator* films, and it brings us back to all those questions of trust. These will be hard issues to negotiate, and they make current debates over AI and whether or how to set limits around it seem almost trivial.

Over the longer haul, it is possible that by somehow bypassing or rewiring that motivational circuitry, Homo Chrononauticus could become more obedient to the future (and past), less interested in surprise and more interested in fulfilling what has been predicted and predetermined. Another story by Chiang, "A Story of Your Life," the basis for the movie *Arrival*, is a lovely examination of precognition and an alien civilization that is able to experience itself in 4D. For the heptapods in that story, seeing the future actually brings with it a sense of urgency to fulfill what is foreseen—something like Nietzsche's *amor fati* or love of (one's own) fate, even when that fate includes tragedy or misfortune. I know a precog who consciously tries to live this way, seeking to build what she calls "congruence" with her future and past self. It is not too unlike Mossbridge's "love" in her vision of a congenial time-traveling future that avoids time wars and other darker possibilities.

Elsewhere I have argued that Predictor-like devices, as troubling as they seem at first, may in fact be useful physical koans, propelling users to an enlightenment state that transcends our current hangups about free will.* Zen satori experiences often are realizations of the underlying permanence of things that profoundly challenges our commonsense notions of dynamism and spontaneous action, but in a blissful and paradoxically liberating way. So I always advise people to just put the whole question of free will on a mental shelf and confront the idea of retrocausation and time-travel within a single self-consistent block universe, and see whether you don't ultimately reach an epiphany that thinking about free will is just a hindrance, holding you back. The block universe is a spiritual riddle that can push you one step closer to the Archimedean perspective of the heptapods in *Arrival*.

Akinetic mutism, in other words, could just be the sullen-teenager phase in our maturation as a species, as we're getting used to the new reality and evolving toward a newer, wiser, timefaring self.

---

* Wargo, *Precognitive Dreamwork*.

3.
The Tesseractuality in the Room:
Wormholes, Timeships, & UFOs

"Instead of looking at the screen, what I want to do is to turn around and look the other way. When we look the other way what we see is a little hole at the top of the wall with some light coming out. That's where I want to go. I want to steal the key to the projectionist's booth, and then, when everybody has gone home, I want to break in."
—Jacques Vallee, radio interview

The story so far: We, here, now are living in the immediate prehistory of our timefaring future. Technologies already on the drawing boards may be capable of sending information into the past and even allowing two-way communication between users at different times. A quantum server that allows chronophone calls and c-mails sent across the fourth dimension could rapidly bootstrap our civilization—or at least, elements of our civilization—into a future with the technology to physically travel across the fourth dimension too.

Although physical visits to other times in past or future history appear occasionally in literature before the 19th century, the idea of traveling in time using some form of technology was purely a Victorian invention. H. G. Wells usually gets the credit, with his 1888 short story, "The Chronic Argonauts" and then his 1895 elaboration of the same premise in *The Time Machine*. But a journalist named Edward Page Mitchell beat Wells to the punch with an anonymous 1881 story in a boys' magazine, "The Clock That Went Backward," about a history student whose great aunt's time-traveling clock brings him to the 1574 Spanish siege of Leiden, where he saves the day using his foreknowledge of how the battle unfolded. As far as I know, Mitchell was not only the first time-machine storyteller but also the first to explore the idea of the future influencing the past without causing paradox—that is, my favorite beast, the time loop.*

Mathematicians and theologians in the Victorian era loved speculating about higher dimensions. Edward A. Abbott's 1888 fantasy, *Flatland*, is about a three-dimensional Sphere who descends into a two-dimensional world, bewildering one of its inhabitants, A. Square, and then giving that Flatlander a journey through his own higher reality. This was obviously a way to jolt readers into realizing that our 3D reality might be a "flat" shadow or cross-section of a higher, 4D world.† A mathematician of the time, Charles Howard Hinton, wrote books on the fourth dimension and taught mental exercises to help people visualize 4D incarnations of ordinary solids.‡ He

---

* See Nahin, *Time Machines*.

† Edwin A. Abbott, *Flatland* (New York: Dover Publications, Inc., 1992).

‡ Charles H. Hinton, *Speculations on the Fourth Dimension* (New York: Dover Publications, Inc., 1980).

coined the term *tesseract* to refer to a 4D cube or "hypercube," although the term was later adopted by science-fiction writers to talk about what came to be called wormholes and similar exotica. But Abbott and Hinton were thinking solely in terms of additional spatial dimensions. It was Wells in *The Time Machine* who explicitly articulated that the fourth dimension was time. With this conceptual leap, the writer beat Einstein and relativity by 10 years.

Einstein's 1905 paper "On the Electrodynamics of Moving Bodies" consecrated the fourth-dimensionality of time for science, and the result of this paradigm shift was the block universe: space and time as a 4D unity or continuum. Travel in a leftward direction across the fourth dimension was not the point, and the possibility of such travel was not immediately apparent to Einstein. In fact, in a 1907 paper on relativity's implications, he wrote that the idea of sending a message backward in time—which the math seemed to indicate was possible if it went faster than light—"conflicts with the character of all our experience to such an extent that this seems sufficient to prove the impossibility of the assumption [of a velocity exceeding light]."

It conflicts with our experience, so no dice. Denial based on a thing simply being unfamiliar is a weirdly common pattern in science, as we've already seen, and even Einstein succumbed occasionally. It was his friend Kurt Gödel's solution to his field equations that eventually convinced him of time travel's real possibility.

We saw in Chapter 1 that wormholes could allow objects or people to travel not only great distances across space but also to points in the future or past. Some astronomers have proposed that naturally occurring wormholes could be plentiful in the universe, but they have never yet been seen. Manufacturing a wormhole from scratch is theoretically possible but is often assumed to be a technological feat well beyond our current capabilities. Michio Kaku claims that to open one and keep it open would require the energy output of an entire galaxy—quite a lot. But then again, Kaku also thinks that people could commit causality violations if they traveled back in time, and we know that's wrong. Also, most proposed solutions to stabilizing a traversable wormhole require the still-hypothetical "negative mass." But physics, like everything, evolves, and there may well be discoveries and developments in our nearer future that could allow us to create such devices. Harnessing so-called zero-point energy, the enormous energy that exists in seemingly empty space, could be part of the solution.

In physics texts, you will still see wormholes depicted like coffee-mug handles or tubes with flared ends, the "wormhole mouths" (or sometimes, "throats"). This has led to the impression, conveyed in a million science-fiction movies, that traversing a wormhole would be like passing through a door or gate, or maybe a sort of whirlpool, zooming through a twisty chute at a water park, and emerging from a similar flat door or gate or whirlpool at the other end. It is an artifact of trying to represent a non-Euclidean object Euclideanly, on graph paper.

Spacetime is often spoken of as a "fabric," which is also misleading but at least gets us a little closer to the reality. In Madeleine L'Engle's 1962 novel *A Wrinkle in Time*, one of a trio of time-traveling spinsters demonstrates the principle of higher-dimensional travel literally by holding out the cloth of her robe with two hands and bringing her hands together, letting the cloth in between drop. Teenage Meg Murry and her companions are asked to picture how an ant at one end of the cloth could quickly get to the other end by stepping over that fold. "Wrinkle in time" (or timewrinkle) never took hold except among young readers—the other term L'Engle used, tesseract, had its origins in Hinton's writing about higher spatial dimensions—but bringing two points in spacetime into convergence or overlap is a better way of picturing wormholes than tubes or tunnels. They are really junctures, places of intersection or convergence.

It was really not until Christopher Nolan's *Interstellar* in 2014 that wormholes were depicted accurately in popular culture. As shown in that film, one of these junctions would appear to an observer as a *convex*, three-dimensional object, most likely a sphere, through which the other location was visible as through glass. The refracted light of stars seen through and beyond it would be interpreted visually by an approaching astronaut as reflected light, like off the surface of a shiny object. As they approached and lost sight of the "rim" in their peripheral vision, the astronaut would see that there was actually no surface and that the convex appearance had been an optical illusion. Rather than sliding through a twisty-bendy chute, the experience would be one of seamless (but incredibly weird) transition from one environment to another.

Since wormholes link two points in space*time*, naturally occurring ones might not be static, just sitting there hanging out, "always open" like your corner convenience store. They might have a finite, brief duration in time, meaning they might fade or grow into being from a tiny point and then disappear again just as quickly after reaching their full diameter. This temporary aspect is one thing Terry Gilliam's 1981 film *Time Bandits* got right: The holes used by the bandits to commit their crimes are localized in time, not just space, and the time travelers have to be at the right place at the right time to travel through them. (Gilliam did commit the usual mistake of making the openings flat and doorway-like, but it's such a wonderful movie that I forgive him.)

Before it becomes possible to open and stabilize wormholes big enough for people or spaceships to pass through, it may be possible to open tiny ones that can be used to send nanotechnology, or ants. Some physicists have proposed that incredibly tiny Einstein-Rosen bridges are what accounts for quantum entanglement—that basically, wormholes might connect everything, on the micro scale. One proposed way of creating a wormhole for human use is by "extracting" it from the micro quantum foam and somehow enlarging it. Julia Mossbridge imagines opening a wormhole just big enough to send a viral vaccine vector into the past, which could then spread and pre-inoculate one's population from some new pandemic threat.*

But there are other options for traveling in time, if you are a person or person-sized.

Timeships

Mitchell's time-traveling clock notwithstanding, Wells' time machine is the first true temporal conveyance in fiction. Vaguely described only as a spindly contraption with a seat and handles, it reminds readers more of a bicycle than anything else, and this in fact may have been deliberate on Wells' part. Bruce David Sommerville has made an interesting case that *The Time Machine* was really a clever "cipher-story" describing an afternoon cycling excursion outside London, not a "real" journey through time. Wells was an avid cyclist, and bicycles were the acme of modernity in his era. Wells may have been using time travel as a metaphor for the speed and freedom offered by these devices.† Which may help explain why Wells didn't do much to make his time machine, or travel through time, particularly realistic.

The anonymous Time Traveler sits on his machine, pushes a lever, and experiences the aeons whipping by like a blur. When he stops in England's far future, he dismounts his contraption and steps into a radically changed land-

---

* Mossbridge, "How to Avoid the Time Wars."

† See Bruce David Sommerville, "The Time Machine: A Chronological and Scientific Revision," *Wellsian* 17 (Winter 1994): 11-29.

scape; his building and laboratory have long since crumbled to dust. Wells implies that the Time Traveler and his machine vanish from the present and reappear at the destination time; it conveniently relieves the author of having to explain how, in those hundreds of thousands of elapsed years, nobody has interfered with him. Really, what an outside observer should see is a motionless statue of a man in a steampunk contraption, unblinking, unmoving, all physical processes in his body slowed to practically a halt, his skin like rock. He would be quite the conversation piece, even an Object of Art, and ultimately an interesting feature in a wilderness landscape.

After his adventures in the future, the Time Traveler returns to his home time, and here logic breaks down again. What would it mean to travel back in time in a stationary time machine? What exactly would "appear" a few minutes or months or millennia ago at the same physical location? Again, in the block universe of Einstein and Minkowski, objects are really worldlines, snaking their way through the block. The time machine's past is just its own history, its worldline. These details didn't matter to Wells, who later admitted he didn't believe real time travel was possible anyway (supporting Sommerville's notion that it might really have been a bike trip).

Let's get a little bit more precise about what travel in a time machine really means: It is the ability for a time traveler to occupy a *reference frame* that is distinct from the surrounding reference frame. To do that, you need to be moving through space. Faster-than-light travel is tantamount to time travel, in fact, since the photons that carry causality through the block universe define what is called the "light cone" of the observer.

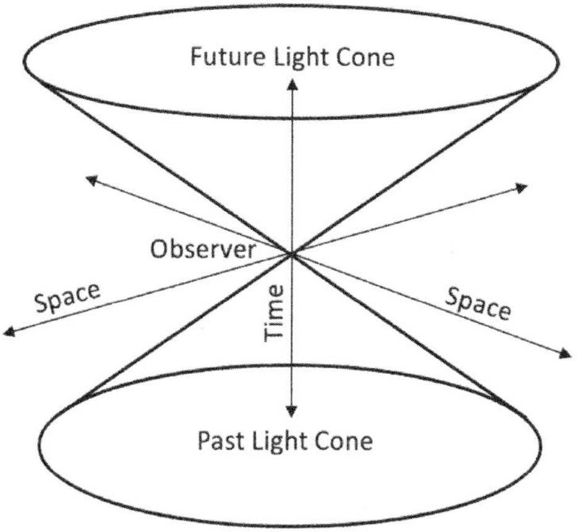

Spacetime Diagram

The cones' "walls" represent the speed of light; everything within the bottom cone is the potentially known past. Everything in the upward-facing cone is the future accessible through conventional or relativistic spaceflight. The "spacelike" realm outside the cones is inaccessible without traveling faster than light.

Traveling faster than light may be possible using a so-called Alcubierre warp drive, which modifies that spacio-temporal fabric or "metric" around the ship such that it surfs a wave of expanded and contracted spacetime. Like an artificial wormhole, it depends on still-hypothetical exotic matter having very specific properties, but otherwise it doesn't violate currently accepted physics. First proposed in a 1994 paper by a theoretical physicist at the University of Wales named Miguel Alcubierre, a ship equipped with such a drive is sometimes touted as getting you to Proxima Centauri, about four light years distant, in just a few weeks or months.* But even more interesting is turning around and coming back from Proxima Centauri—depending on your exact speed, you might arrive about 8 years before you left.

Miguel Alcubierre himself, in trying to evade the causal conundrums posed by his craft, invoked Stephen Hawking's fantastical chronology protection conjecture: Something might go haywire to prevent you from using a warp drive to travel backward in time.† But this is just part of the larger hysterical reaction against supposed causality violations that we dispelled in previous chapters. The arrival back on Earth of the Proxima Centauri mission 8 years before its departure (and the fact that slightly older and fantastically more experienced clones of the crew trainees walk among them—see below) is perfectly okay; it would be part of the backstory, the secret history perhaps, of their society and of their journey. This is why I said our timefaring future will get really, really weird. We need to stretch our minds to encompass a brand-new set of possibilities, not just new physical possibilities but unprec-

---

* Miguel Alcubierre, "The Warp Drive: Hyper-fast Travel within General Relativity," *Classical and Quantum Gravity* 11(1994): L73-L77.

† Miguel Alcubierre, "Faster than Light" [slideshow], Internet Archive, retrieved December 15, 2024, https://web.archive.org/web/20160318223348/http://ccrg.rit.edu/files/FasterThanLight.pdf.

edented social and cultural possibilities, and not continue to hide behind flimsy cardboard conjectures.

Zooming to light speed and beyond may not even be necessary for a time machine. Another way to travel in time is to bend your worldline back upon itself, making an about face. In a 2013 paper cleverly called "*T*raversible *A*chronal *R*etrograde *D*omains *I*n *S*pacetime," physicists Benjamin K. Tippett and David Tsang proposed that a true time machine like Doctor Who's TARDIS could be created by isolating a bubble of spacetime geometry from its surrounding temporal river. Conceivably, a vessel in such a bubble could reverse its trajectory along the block universe's time-axis. This would be a true, fully functional time machine, or what I like to call a *timeship*.*

Exactly how to build such a time-reversible vessel remains as theoretical as building a wormhole or an Alcubierre warp drive. But essentially, by remolding spacetime using fancy metamaterials and/or other tricks, the timeship's bubble would insulate the thermodynamic reference frame of a timeship pilot and her cat from the reference frame of the surrounding environment. Even traveling conventionally "forward" in time, they would be unaffected by the G-forces that would affect the occupants of a conventional plane or space capsule—true also of an

* Benjamin K. Tippett, David Tsang, "Traversable Achronal Retrograde Domains In Spacetime," arXiv:1310.7985v2 [gr-qc], November 4, 2013, https://arxiv.org/pdf/1310.7985. A more accessible explanation can be found here: Benjamin K. Tippett, David Tsang, "The Blue Box White Paper," arXiv:1310.7983v1 [physics.pop-ph], October 29, 2013, https://arxiv.org/pdf/1310.7983.

Alcubierre warp ship. But even more cool is that the chrononaut could pull a lever and reverse the flow of time within her timeship's hull relative to that of the outside flow of events—effectively making a U-turn in time. The chrononaut would then be traveling backward in time while still experiencing her own time moving forward.

Such a switchback would look very funny to an outside observer: It would really seem to them that there were *two identical timeships colliding and immediately vanishing*. The two objects are/were really the same vessel, one in the normal reference frame and the other traveling into the past. "They" vanish because the timeship has done a U-turn and left the local reference frame altogether. (Animations of this and other possibilities described in this section can be found on the "Timeship Motion Signatures" page of my website, ericwargo.com.*)

Quite distinctively, time travel creates doppel-gangers—clones—in violation of the conservation laws physicists used to preach. It's also true of wormholes. And it is why a timeship has to travel in space too—otherwise, its switchback would "conflict with itself" when it makes its temporal U-turn.

After the chrononaut's destination in the past is reached, she could remove her sleeping cat from the warm console and pull the lever again, rejoining the local reference frame. To an outside observer (at an earlier time), that second turnabout would be the opposite of the previous situation: It would look like an object (the timeship)

---

* URL: https://ericwargo.com/timeship-motion-signatures/

expanding into existence from nothing and immediately splitting into twin objects traveling apart.

Putting it all together, a basic S-shaped switchback in time for a single timeship would look like twin timeships popping into existence out of nowhere and one of those ships colliding with a third ship and disappearing. At any given time, a timeship always appears singly or in multiples of three, because in addition to the backward-traveling ship, there are two that are moving forward—the timeship before the turnabout and the same timeship after the second switchback.*

And there would be nothing preventing the chrononaut from switching the temporal vector of her timeship repeatedly, perhaps to get multiple vantage points on the same external event. A single time-reversible object can quickly look like a swarm of identical objects (always an odd number) dividing and merging in a hypnotically beguiling pattern.

I assumed for simplicity's sake that a time-reversed object looks to an outside observer the same way it does when it is not time-reversed. But depending on how the

---

* The object-multiplying effect of time travel was depicted, incidentally, in the beyond-wonderful film *Yellow Submarine*. After accidentally traveling backwards through the Sea of Time, the Fab Four reverse course and then observe another version of themselves traveling backwards. Then they sing "When I'm Sixty-Four," and yours truly gets a little tear in his eye ... just before he gets all pedantic and mansplains to bored-looking wife and daughters that there ACTUALLY should be a third submarine, the original version of themselves before the switchback. When I said doppelgangers earlier, I should have said trippelgangers.

trick is achieved, the hull of the timeship in reverse might not be able to interact with the surrounding environment electromagnetically—in other words, it might de facto be cloaked. If that's the case, then a simple brief time reversal followed by switchback could look like a duplicate timeship appearing at a distance from the first, followed at some point thereafter by the disappearance of the first object. However, even if the hull itself is invisible when time-reversed, the bubble-like distortion around it might still be visible, so a witness might still see something, a jelly-like disturbance perhaps.

A more crucial problem is: What would our chrononaut see on her timeship's viewscreen? Relative to ship time, everything outside the timeship's bubble should run in reverse—we could imagine it like the film of history being shown backwards. But in fact, the chrononaut might not be able to see it. Traveling backward against the external temporal reference frame, she would be effectively blind to the world outside, since the photons needed for vision also will be receding toward rather than emanating from their sources. In other words, everything outside the ship might look black from the pilot's vantage point during a retro-jaunt. It could be just one of many factors making temporal navigation difficult (we'll discuss others in the next chapter).

There could be a workaround here, however. In the late 1960s, physicist Gerald Feinberg proposed the existence of faster-than-light particles called *tachyons*, which would travel in temporal retrograde. Thus far, we still don't have evidence of a tachyon realm—our eyes certainly can't see such particles. But if a tachyon spectrum does exist, presumably the builders of a timeship also could make a special viewscreen (or goggles) to translate tachyons into visible light, enabling our chrononaut to enjoy the "world-rewinding" experience of timeship travel against the local temporal current.

Real Relativity

Tachyons make a nice segue to the next part of this chapter, as they bridge the world of science and the paranormal.

After they were first proposed, tachyons seemed to offer a possible explanation for a range of paranormal phenomena including ESP. It was research on tachyons that led physicist Hal Puthoff, who now works on the problem of spacetime manipulation and exotic propulsion, to intersect with psychic Ingo Swann and create a CIA-funded ESP research program at Stanford Research Institute. Their research led to the Pentagon's storied Stargate remote-viewing program. After reading a 1974 article about tachyons by Arthur Koestler in *Harper's*, Philip K. Dick also took interest in these hypothetical particles as an explanation for his own precognitive experiences in dreams and in his writing.*

Tachyons as an explanation for precognition makes intuitive sense ... until you consider that such particles would be about as useful for directly seeing into the future as photons are for directly seeing into the past—that is, not at all. Looking through tachyon goggles, the world would

---

* See Wargo, *From Nowhere*, for more on this.

look just like it does normally. The exception is astronomy: With very distant objects, it does make a difference whether you are seeing the past of an object (as we do with photons—nanoseconds in the past for anything in our immediate environment)—or the future of an object, as we would when observing in the tachyon spectrum.

Through a tachyon telescope, nearby stars would appear more or less in their familiar positions, but more distant objects would be offset, and distant galaxies would be in very different positions and configurations. Tachyon telescopes of the 22nd century might observe much more evidence of sidereal engineering in our own and neighboring galaxies, since the future may hold much more time for civilizations to evolve and reshape their stellar neighborhoods. A tachyon telescope may reveal lots of Dyson spheres and ringworlds, for instance. And who knows what kind of really massive galactic engineering might be visible in the farther distance. With a powerful enough tachyon telescope, we also might be able to see to the literal end of the universe, rather than its beginning.

With all this talk of saucers and objects popping into and out of existence, the word "UFO" might have come into your mind. (In my household, we say "UFO," not the newer, blander term "UAP.") So, let's address the elephant in the room. If humans in the future will be able to build timeships, then we should look for evidence of it now, and the UFO phenomenon is certainly the most obvious place to look.

UFO's are undoubtedly a controversial idea, although nowhere near as controversial now as they were in previous decades. It is widely known that military pilots see and interact with anomalous objects routinely; and government whistleblowers just in the past few years have testified before congress that there are very secret Pentagon programs involved in studying them and perhaps even retrieving crashed vehicles. No longer can these phenomena be explained away as delusion or swamp gas or the planet Venus, as they were in previous generations. The suspicions that there could be mundane explanations like secret human technology (perhaps of our global adversaries), as well as mind-control experiments along the lines of the CIA's infamous project MKULTRA, have not abated; yet a core of UFO cases seems to defy what present-day humans, even with enormous technological advantages, could achieve.

The idea that UFOs could represent time-traveling technology rather than space-traveling technology is not new. It is implicit in some of Jacques Vallee's classic writings on the subject. For example, in his 1969 book, *Passport to Magonia*, the French astronomer suggested that the UFO phenomenon is some kind of active shaping of our imagination possibly projected from some interdimensional elsewhere or elsewhen. Of one Libyan encounter in 1954, he wrote: "If it were possible to make three-dimensional holograms with mass, and to project them through time, I would say this is what the farmer saw."*

---

* Jacques Vallee, *Passport to Magonia* (Chicago, IL: Regnery, 1969), 152.

You may recall from Chapter 2 that darker possibilities of history gardening include husbanding of other groups (or species) in ways that will benefit your own group later. Such husbanding would be an interesting interpretation of Vallee's "control system" hypothesis: Past and present-day humans may be being husbanded by some other future civilization for their benefit. ("Is human imagination alone capable of playing such tricks on itself?," the astronomer asks. "Or should we hypothesize that an advanced race somewhere in the universe and sometime in the future has been showing us three-dimensional space operas for the last two thousand years, in an attempt to guide our civilization?"*) Or, these phenomena could just be future us, husbanding ourselves.

The idea that ufonauts could be our own descendants has gained renewed cachet over the last few years, thanks in large part to the work of Montana Tech anthropologist Michael P. Masters. In his comprehensively argued 2019 book *Identified Flying Objects*, and then a 2022 sequel, *The Extratempestrial Model*, Masters draws on several lines of evidence to support the greater plausibility of a time-travel explanation for the UFO phenomenon than a space-travel one. The humanoid form of most reported visitors and ufonauts is a natural starting point.† Masters agrees with the oft-expressed argument that real ETs would be unlikely to look anything like us, given the wide variety of planets

---

* Ibid., 153.

† Michael P. Masters, *Identified Flying Objects* (Butte, MT: Michael P. Masters, 2019); Michael P. Masters, *The Extratempestrial Hypothesis* (Butte, MT, Full Circle Press, 2022).

and habitats where life could emerge throughout the universe. A *Star Trek* (or *-Wars*)-style cosmos filled with intelligent beings having the same basic body plan simply reflects a kind of anthropocentric prejudice and the exigencies of cheaply depicting aliens in pre-CGI Hollywood, according to this view.

Personally, I disagree with that part of Masters' argument. Being highly reliant on technology is a very specific kind of niche, one that favors a certain body plan: upright posture leaving something like hands free to manipulate objects, and so on—as well as a particular evolutionary backstory with specific constraints leading to that body plan (the hands would have evolved from forefeet, etc.). Convergent evolution might thus tend to produce spacefaring civilizations that are humanoid or were once humanoid at some point in their past. I would not be surprised if humanoid turns out to be a common if not universal spacefaring body plan (or phenotype) across the universe.*

However, where Masters' argument becomes most powerful is when he addresses the specific anatomical traits of reported visitors. Commonly described features like small stature, enlarged crania, and large eyes conform very well with hominin evolutionary trajectories he studies as a physical anthropologist. Central to his reasoning is the principle of *paedomorphosis*, retention in adulthood of juvenile traits of a species' ancestors. We did not evolve

---

* Eric Wargo, "Humans Everywhere (the REALLY Anthropic Cosmos)," *The Nightshirt*, October 24, 2013, https://www.thenightshirt.com/?p=1535.

from chimps, but we do share with them a very chimp-like common ancestor, and chimps' kids' faces have proportions remarkably like those of adult humans. By the same token, the resemblance of many reported "ETs" to human children would make perfect sense if these beings are really our descendants.

I'm in no position to evaluate the quality of the close-encounter and abduction evidence that Masters draws on for his argument—some of the cases (not all) strike me as more likely the result of complex human deception and mind-control experiments. But an additional interesting line of evidence comes from sighting reports that could support the existence of time-traveling technology operating in Earth's atmosphere. The impossible speeds and maneuverability commonly reported are suggestive of something like the "metric engineering" that Puthoff studies and that could be involved in timeship construction. Physicist Jack Sarfatti thinks the hulls of the now-famous "Tic Tacs" encountered by Navy pilots stationed on the USS *Nimitz* in 2004 are made from layers of molecularly engineered materials able to slow down the speed of light, thereby insulating the occupants from the local gravitational field, akin to an Alcubierre warp drive (but in this case, still traveling slower than light).* He has also

---

* Jack Sarfatti, "How Flying Saucers/Tic Tacs Really Work V12" [unpublished conference paper], Academia.edu, retrieved December 15, 2024, https://www.academia.edu/40756596/How_Flying_Saucers_Tic_Tacs_Really_Work_V12.

speculated that these objects are indeed time machines, made by us, in our future.*

Masters also notes that many UFOs seem to disappear and/or teleport, which is highly suggestive of some kind of timeship behavior. Especially intriguing to me are reports that sometimes appear in the UFO literature of lights or solid objects that behave suggestively like the timeships described earlier: confusingly splitting into two or more objects after appearing and/or converging, merging, and vanishing. Again, splitting and merging is a motion signature consistent with a single timeship reversing its direction along the T-axis of local time. Also, UFOs sometimes appear in swarms, which is what a single timeship scrubbing back and forth in time to gain multiple vantage points on an event could look like.

Even "fleets" or "formations" of UFOs (as long as they are in odd numbers) could potentially be a single timeship performing complex temporal maneuvers. Wernher von Braun is alleged to have recorded in his diary that a swarm of luminous orbs had been seen by a witness or witnesses during the Trinity nuclear test in Alamogordo, New Mexico on July 16, 1945.† (The Nazi scientist was brought to Los Alamos as part of Project Paperclip a few months

---

* Simon Greene, "Top physicist claims USS Nimitz UFO 'time travelled' to reach 19,000mph," *Daily Star*, March 20, 2020, https://www.dailystar.co.uk/news/weird-news/top-physicist-claims-uss-nimitz-21726921.

† "Dr. Pippa Malmgren on The Space Race, Geopolitics and the Quest for Ultraterrestrial Life," YouTube, https://www.youtube.com/watch?v=XvsTHj-MNAc.

after that test.) Could it have been a single timeship observing that historic event from multiple angles? All we can do is speculate. But if anything comes through in many UFO reports, it is that these objects (like everything) have extension in time, and that what we are seeing could be merely a cross-section of a highly complex, 4D reality, a "tesseractuality."

What about wormhole travel? There's a now-famous description of an aerial portal sighted at the controversial paranormal hotspot Skinwalker Ranch in Utah, which the witnesses described as looking "like a three-dimensional orange tunnel that receded away from them."* This sounds to me like either a sky hologram meant to fool people or else a rancher concocting a story based on popular representations of wormholes in books and movies. For reasons discussed earlier, wormholes aren't tunnels. However, there are many reports in the UFO literature and in older accounts of the unexplained in which witnesses saw what looked like other worlds or distant cities through openings in the sky. Some of these sightings could represent some geomagnetic phenomenon like "sprites" (which look a bit like spired glowing cities in the upper atmosphere), but who knows what else could be possible.

And here's the thing: Assuming that future humans eventually surmount the technical challenges of opening and stabilizing a wormhole, there's no need to limit wormhole travel to conventional air- or spacecraft trav-

---

* Colm A. Kelleher, George Knapp, *The Hunt for the The Skinwalker* (New York: Paraview Pocket Books, 2005), 64.

eling *through* wormholes. A more interesting possibility is that some UFOs could themselves *be* wormholes. Time-farers could combine wormhole technology with a TARDIS-style vessel carrying the wormhole "mouth," and the result would be remarkably similar to what is reported by some people who have had truly bizarre, very close encounters with UFOs and their occupants.

Because it was a wormhole linking two deep-space points, Nolan's wormhole in *Interstellar* looked from a distance a lot like a glass Christmas tree ornament. A wormhole created in the high-tech laboratory of some future Wormhole Inc. and projected or carried to our time would look very different: Instead of stars, what a witness would see are the insides of the machine or chamber generating the wormhole, turned inside-out.

From a safe distance, an observer would interpret its surface as convex (pushing toward them). If the chamber was metallic, perhaps like a big vacuum chamber in a physics lab or the inside of a fusion reactor, it would look from a distance visually solid and round, or perhaps squatly cylindrical or even saucer shaped, depending on the shape of the chamber. If there were portals or windows on the chamber walls with technicians outside looking into it, these would appear from a distance like portals on the craft's exterior with ufonauts inside, looking out.

Only as you approached this object—or as it approached you—would its perspective seem to distort and your brain might finally grasp you were seeing into the insides of a machine or chamber of some sort—a seamless transition from outside to inside. Although more likely, you, a 21st-century human, would be too confused, frightened, and upset by the encounter to grasp anything at all. We simply have no familiar experience to serve as a touchstone for such a thing. But abductees often describe interiors that are bigger on the inside than the outside or that don't correspond well to the tightly packed and neatly arranged environment one would expect of a spacecraft. It has frequently been suspected that the examinations (if real) are actually taking place at an Earth-bound facility somewhere, not on the UFO. If a "ship" was really a laboratory wormhole chamber projected to another locale, it would help explain these kinds of details.

Another problem that the UFO-as-wormhole possibility would help solve is the incredible energies that are sometimes estimated to be associated with UFOs. In a 1998 paper, Vallee estimated the luminous power output of six UFOs in his encounter files, finding values potentially up to 30,000 megawatts in one case, which is not even including the power necessary for propulsion.* If some of these objects were really wormholes carried from somewhere or somewhen else, instead of packing massive power plants within their visible dimensions, you could have a building- or city- (or planet-) sized power plant off-stage supplying all of the energy. Again, what looks like a UFO could really just be the Wormhole Inc. UFO-shaped chamber turned inside out, so the chamber itself need not contain anything—no people, and no power plant. (Even the timeship or drone carrying the remote wormhole mouth

---

* Jacques F. Vallee, "Estimates of Optical Power Output in Six Cases of Unexplained Aerial Objects with Defined Luminosity Characteristics," *Journal of Scientific Exploration* 12(3) (1998), 345-358.

from place to place could perhaps be powered by that distant or future power plant, via a cable or some form of wireless power transmission passing through it.)

A wormhole generator also solves the more fundamental "relativistic" problem: How to bring people (e.g., relatives) to you instead of going to them. Because, as grown-ups realize on occasions like Thanksgiving, *it's more fun being a host than a guest*. A wormhole enables you to interact with distant objects and people from the comfort of your own laboratory, not have to pack yourself into a cramped ship and go on an annoying or dangerous journey. So, we should bear in mind that interacting with or manipulating *unwitting* guests is probably the least of a wormhole generator's many fabulous uses.

The Dark Cone

In an interview, Vallee expressed a desire to break into the projection room at the back of the human theater and inspect the mechanism projecting UFO anomalies on our consciousness and radar screens.* I tend to agree with Masters that we will break into that room only to find ourselves not only taking over the controls of the projector but actually inventing and installing the projector in the first place. The UFOs, we will realize at that moment, were us all along. But the new frontier we will be set to navigate

---

* "Brother Jacques Vallée: Excerpts from an interview with Jacques Vallée," retrieved January 9, 2025, https://www.bibliotecapleyades.net/bb/vallee.htm.

(the past) will turn out to be surprisingly unknown, surprisingly full of mystery, despite the fact that we have been there already.

One of the benefits of thinking about time travel is that it makes us humble about our supposed knowledge … of anything. The past—the bottom-facing cone of the space-time diagram on page 101—is all-too-readily imagined as the clear light of a lamp shining downward on past history, subtly giving the impression that the past is a *known* past. But we should really think of it more like a rubble pile, with a few illuminated patches here and there over a vast, dark, unknown, unilluminated mass. Think of how vague and spotty your own memory is and how little of your life is actually recorded and thus reviewable in any way—even now, in our age of smartphones—and then project back into a history that lacked even basic recording technology. We must set aside our pride and arrogance in how much we think we know about our past and present. The light cone is really a *dark* cone—in both directions.

Stephen Hawking concluded his chronology-protection-conjecture article flippantly by saying he had proof that time travel was not a thing: the fact that our world isn't full of time travelers from the future, visiting us. Masters points out how myopic and presumptuous this is, comparing it to Enrico Fermi's famous "where are they?" at Los Alamos in the 1950s. At the time Fermi made his remark implying that we ought to have evidence of other civilizations but don't, flying saucers had been buzzing the planet for several years, luminous orbs were seen frequently around Los Alamos, and so on. Which is not to say that those saucers and orbs were extraterrestrial vehicles—I personally doubt that—just that there is way more in heaven and earth than is dreamed of in our arts and sciences, as anybody paying attention can attest.

It takes us back to the theme we started with, of the imagination. Unless and until people have a realistic picture in their heads of time travel and time travelers, these things will not be part of anybody's historical imagination. Where are all the time travelers at the Battle of Waterloo or the D-Day Invasion, for instance? The history books don't tell us of any, but who was even looking for them? Who would have even known what to look for? For that matter, where are they in the Gospels? Shouldn't the Sermon on the Mount have been crowded with them? Or the Crucifixion? The New Testament is mum on time travelers.

Lo and behold, the Gospels are *not* mum on mysterious angels appearing to Mary and announcing her immaculate pregnancy, or on stars appearing over Bethlehem right when Jesus is about to be born, and mysteriously clad strangers "from the Orient" showing up for the event, etc. And we actually have no idea who was there on the Mount when Jesus gave his famous Sermon—or for that matter, who most of his followers and apostles were. They could *all* have been time travelers, for all we know. His whole audience and following might have been Christians arrived from the far future, disguised as local shepherds, for the amazing opportunity to hear their Savior in person … or perhaps, see him die. In his novel *Up the Line*, Robert Silverberg imagined "an audience of billions of time travelers piled up in the past to witness the Crucifixion."*

If you had a "duh" moment just now, realizing that time travelers could already be all over the place within our historical narratives and myths, not "just appearing now" when we thought of the possibility, then it is a symptom of still thinking linearly about time in a universe where loops rule. It's the same fallacy as Hawking's mad scientist. It can't be stressed enough that history doesn't change all the sudden, in the present, when it belatedly occurs to us that someone, or many people, could visit the past. If such visits occurred—and why not?—then they are a big and thus far unaccounted-for part of our backstory. The only thing that changes once we acknowledge the possible reality of time travel is our understanding, our interpretation, what we presumptuously call our "knowledge" of what might have really happened in the past.

Everything we think we know about the past has to fit within our current assumptions, and until almost the present historical moment, the very possibility of time travel has not been taken seriously by anybody. I'm not saying those were time travelers at the Sermon on the Mount or watching the Crucifixion—for reasons we will soon explore, I seriously doubt they were—or even that the Gospel stories reflect real events. But we just can't assume

---

* Quoted in Nahin, *Time Machines*, 280. Jesus himself has often been imagined as a time traveler. Michael Moorcock wrote a novel, *Behold the Man*, about a time traveler going back to first century Galilee and finding no Jesus, so living Jesus's life himself (including dying on the Cross).

that what we "know" about the past is anything more than a tapestry of beliefs, assumptions, and interpretations, based on current understandings derived mostly from fallible authorities with their own limited perspectives and biases, and always subject to revision and updating as we gain new information.

There is no reason the landscape of the past—that rubble-pile-like "dark cone" of our past—might not already contain technology, information, and causal arrows that have looped back from the future. I can look at the clutter on my desk and not really "know" that familiar objects in my environment are not cloaked time traveling drones from the 27th century, for example. Such speculation begins to resemble the fictional writings and authentic delusions of one of the most interesting thinkers on the subject of time and its swerves, Philip K. Dick. In many ways, the world we are entering, our timefaring future, will be PhilDickian through and through. But it will not be PhilDickian in the way most people imagine, some maze of alternate histories and time-forking realities. It is its too-solid, too-singular *actuality* that will threaten to drive us bonkers, unless we do some heavy meditation to prepare. (We'll come back to this "radical actuality" in the last chapter.)

Besides the acquisition of precious material resources and planting social or technological seeds that will germinate in their own time, gaining new information, shining even a little light into the crevices of that rubble pile, will prove to be one of the major motivations for time travelers. As we will see in the next chapter, "ancient history" may in fact be a priceless commodity in our spacefaring and timefaring future. And this could be one inspiring answer to the question that was forming in your head when I disillusioned you about *changing* the past way back in Chapter 1: What's the point of time travel if you can't undo some personal tragedy or assassinate the baby Hitler?

In Ted Chiang's short story, "The Merchant and the Alchemist's Gate" (another in his collection, *Exhalation*), a Baghdad merchant hires the use of a time portal to go back in time to revisit his wife, who died in a mosque collapse shortly after they had parted in anger two decades earlier. He cannot reach her in time to intervene or rescue her—kudos to Chiang for making his universe self-consistent—but the merchant learns from her nurse that her last words were of her love for him. It cures the merchant of a life of doubt and regret. Thus, through a very human narrative, it reveals one of the main purposes of intentional time travel: the gaining of knowledge for a better tomorrow. Such knowledge can be immensely useful. Sometimes it can be redemptive.

The necessary predecessor to becoming timefaring will be a culture in which the past is seen as a landscape to be explored and interacted with in ways that reveal or disclose new knowledge and reveal new possibilities. Such a culture will not presume that it really knows anything about the past and will have the guiding presumption that its "knowledge"—of anything—is probably wildly wrong.

4.
The Spacelike Calls to Me:
The Challenges of Chrononavigation

If you don't live in a house with young girls and have not seen Disney's *Moana* about fifty times, you may not know about the most epic adventure of human prehistory. So, sit right back and you'll hear a tale …

Starting about 3,500 years ago, explorers from the islands northeast of what is now Papua New Guinea set off on very long ocean voyages eastward, leaving settlements on every island they reached and always forging farther into the unknown. Over the span of a few thousand years, they populated virtually all the widely scattered islands within a vast triangle formed by Hawaii in the north, New Zealand in the southwest, and Easter Island in the east—a sprawling superculture connected by an oceanic superhighway, engaging in trade and ongoing exchange of people and lore, speaking closely related languages, and sharing much the same mythology centered on the hook-wielding trickster Dwayne "The Rock" Johnson. There is evidence that they made it to the Americas and mingled genetically and culturally with populations there as well.

It is impressive for many reasons. The distances in the Pacific Ocean are huge, first of all. Some of these journeys, in single- or double-hulled canoes hewn from tree trunks and lashed together with coconut husk fibers, would have rivaled the famously fraught and uncertain Norse and later Spanish and English voyages to the New World. The problem of wayfinding, when your target destination is tiny and the ocean vast, makes maintaining an active culture of travel and trade between the myriad scattered islands all the more incredible. By and large, Polynesian navigators knew right where they were and where they were going. They could direct their vessels to precise points on a huge, curved triangle of water and keep to a tight, linear course. They did this thousands of years before longitude could be reckoned with Europeans' mechanical clocks.

How the Polynesians achieved this feat is one of those miracles of human ingenuity and cultural transmission that still excites anthropologists. As one of Lin-Manuel Miranda's catchy songs in *Moana* relates, their navigators read the wind and the sky, knew the names of every star, and passed on their elders' lore, generation to generation, in a neverending chain … at least, until European contact. With the adoption of modern, technological navigation techniques, the traditional wayfinding secrets were largely abandoned and eventually forgotten. But in the 1970s, with the help of anthropologists, apprentice Polynesian sailors supplementing the tidbits still handed down by their elders with "frog DNA" of modern research to reconstruct the ancient methods.

One of them was a young Hawaiian man named Nainoa Thompson, who spent months working in Honolulu's Bishop Museum Planetarium to memorize the celestial sphere in great detail and reconstruct methods of determining direction using pairs of familiar stars that form pointers in the sky.* Latitude was reckoned by his ancestors using detailed knowledge of stars known to pass directly over different islands in their vast archipelago, along with many other tools. This incredibly sophisticated celestial navigation was supplemented with a minute understanding of waves and how they break around land masses, water color and what it reveals about depth, patterns of bird flight over open water, and myriad other indicators of position and course available both in daylight and at night. Without LORAN, GPS, or any other modern technologies, they could find their way very precisely to distant, tiny destinations. Between 1976 and 1980, Thompson piloted a traditional sailing vessel on several tight, linear voyages between Hawaii and Tahiti—about 6,000-miles, round trip—using nothing but these ancient methods.

The Polynesian settlement of the Pacific served University of Hawaii anthropologist Ben R. Finney and Los Alamos astrophysicist Eric M. Jones as a model for what our spacefaring future might look like. Their 1985 edited collection, *Interstellar Migration and the Human Experience*, is a priceless relic of a much less dystopian, Reagan-era retro-future that still looked outward to the

---

* Will Kyselka, *An Ocean in Mind* (Honolulu, HI: University of Honolulu Press, 1987).

stars. Taking a social-scientific take on humanity's eventual expansion through our stellar neighborhood and ultimately through the galaxy, Finney and Jones' vision was one of sublight colony ships making long uncertain voyages across vast space. One of the chapters, by astrophysicist David R. Criswell, imagined widely isolated pockets of humanity periodically exchanging the most valuable interstellar commodity, ancient history, via vessels shuttling at relativistic velocities between them.*

The difficulties and perils of navigating the seas of space have always been a favorite theme for the writers of space operas. In order to justify the need for a super-valuable mind-expanding hallucinogen, Frank Herbert outlawed navigational aids like computers in his *Dune* universe due to an ancient restriction on artificial intelligence. But in the absence of such a prohibition, steering spaceships to points in interplanetary or interstellar space is pretty basic stuff, by comparison to what the Polynesians had to accomplish. Unlike distant islands hidden by the Earth's curvature, you can generally see your destination in deep space, and telescopes will make it easy to locate smaller and smaller objectives like planets, so long as there are computers that can help find the optimal course, determining where your target destination is going to be when you arrive versus where it appears now, and so on. Even sci-fi hazards like asteroid fields and nebulae are pretty tame, in reality. Asteroids in

---

* Ben R. Finney, Eric M. Jones, *Interstellar Migration and the Human Experience* (Berkeley, CA: University of California Press, 1985).

the asteroid belt are so spread out that you could travel days without even seeing one. And if you were in a nebula, you'd hardly know it except for the pretty haze around the stars — it wouldn't be like the smoke-and-lightning stuff of *Star Trek II: The Wrath of Khan*.

Finney and Jones were writing in the shadow of Carl Sagan's inspiring but solidly evidence-based PBS series *Cosmos*, when serious futurists wouldn't "go there" in contemplating techniques like antigravity or faster-than-light travel. But it is becoming clearer that our understanding of spacetime is not complete, and that decades- or centuries-long sublight journeys using fusion rockets like the British Interplanetary Society's project Daedalus or the so-called Bussard ramjets featured in Sagan's series just are not, cannot be, the real ticket to interstellar exploration and human expansion. As mentioned earlier, there are hints that U.S. aerospace companies and perhaps other governments have been working on advanced, gravity-defying propulsion technologies since World War II or earlier, and even that basic physics knowledge may have deliberately been hidden from academic physicists—either because the real physics of spacetime engineering are too profitable for aerospace or because they are too dangerous for humanity, or both.* The thing is, as we saw in the last chapter, once you build a ship that can master gravity and/or travel faster than light, you have effectively built a time machine in the bargain.

---

* Jesse Michels, "The Physics of UFOs: Eric Weinstein + Hal Puthoff," YouTube, Feb 11, 2022, https://www.youtube.com/watch?v=iQOibpIDx-4

Recall our Proxima Centauri mission that arrives back on earth 8 years before it leaves. It will quickly become obvious that dumb old Proxima Centauri is less exciting a destination than 8 years ago, if the latter allows you to place some non-obvious bets on the stock market and all the other things that time travel can gain you. Or for that matter, just two years ago … or two months ago, or yesterday. The reality is, time travel may not be the "final" frontier for our civilization, as one might suppose given recent advances in private space exploration. It may really be the next frontier, with interstellar voyaging being a side effect of this technology, not the main objective (we'll come back to why, in the next chapter).

But there's a major problem: For time travelers, every destination in the past, even nearby ones, are over the horizon—literally. They are outside the bounds of visibility and linear causality delineated by that light cone of spacetime diagrams. Physicists speak a funny language and refer to everything accessible through ordinary relativistic physics—i.e., within that cone—as "timelike." Everything in the dark, where no photons come from, is "spacelike." Time travel is to navigate the unseen realm of the spacelike.

Like the line where the sky meets the sea that calls to Moana, the horizon of the light cone beckons. But how will we find our way, once our timeships venture over it?

Bend Your Cone

As we've seen in previous chapters, the topology of the fourth dimension is contoured by possibilities and actualities, the same way the topology of space is—they are all parts of the same continuum. We just call these contours something different when they have to do with history and biography, because we harbor deep-seated, probably ineradicable notions such as free will that make the landscape of our fate look looser, a playground of possibilities rather than something solid ahead of us. But humanity will outgrow these illusions, given world enough, and time.

The real problem for our timefaring descendants will be the problem Moana faces in her little outrigger canoe, finding an invisible destination. Navigating in time presents some of the same challenges as navigating in space, as well as ones that are largely unforeseen, even in science fiction. Novelists and scriptwriters usually fast-forward over the problem of temporal navigation by equipping their time machines with controls enabling the chrononaut to dial in the year and date (in addition to spatial coordinates), so they can easily find the event they are looking for. The most writers will generally do is make some slight nod to imprecision. When jumping through the time portal in "The City on the Edge of Forever," Kirk and Spock overshoot their target time and arrive in Depression-era New York several days before their crazed quarry, Dr. McCoy … But what's a few days when you are a starship captain from the future and want to court the owner of the local soup kitchen? (As we'll see shortly, time travel gives "dating" a whole new meaning.)

In reality, finding a specific target date in time, let alone a specific known or suspected historical event, will be immensely harder than most writers have imagined. A moment's thought should tell you that the fabric of spacetime—that "metric" that Hal Puthoff talks about—doesn't come imprinted with date stamps, or a clock (or a graph-paper-like grid). So, how would our chrononaut in her timeship find her way to a specific destination in past time?

The first principle of timefaring is that to travel in time means to travel in space—possibly, very rapidly. Depending on how fast you have to go to "bend your light cone back on itself," it could mean getting well out of the way of stuff like atmosphere and planets themselves and venturing into deep space. That would certainly be the case if using something like an Alcubierre warp drive. Depending on how far into the past you intend going, you may be able to travel some wide arc over the ecliptic and use the Earth's rotation (through a telescope) as an objective measure of days or its revolutions around the sun as a measure of years. This would depend on somehow seeing in that spacelike gloom, where photons are always fleeing away from your eyes rather than toward them. For the purposes of this chapter, we'll imagine that tachyons can substitute for photons in that realm and that timeships come equipped with tachyon telescopes for their retro-jaunts in time. We'll also assume that the fisheye distortion of rapid travel—everything in your sphere of vision would really appear crowded in front of you as you approached lightspeed—could be compensated for.

If your timeship can make the kind of temporal U-turn described in Chapter 3, high velocities—and thus, forays into deep space—might not be needed, even though you still must be moving in order not to conflict with yourself during the switchback. But you can't just hover for days or years in one spot while you wait for it to be the right time to throw your vessel back into forward. You'd need to be moving, need to land, probably hide out.

If you were a human from the not-too-far future and wanted to visit an event during the modern, industrial era—say, the Trinity test at Alamogordo, New Mexico on July 16, 1945—you might have little trouble finding your target, since the locations are precisely known, findable on maps, and so on. Through trial and error (and luck) you could home in on a precise date through repeated "landings" near populated areas to look at the dates on the covers of newspapers. Which of course presumes that you can speak or at least read the local language. AI will certainly be helpful with this—perhaps a small droid or drone of some sort that can surreptitiously reconnoiter on your behalf. In some cases, even just having a simple radio or TV receiver would do the trick, since you could quickly gather the date from those broadcasts.

For anything prior to the industrial age, though, an astronomical computer will be essential: telling you your date based on the rising and setting and relative positions of heavenly bodies from specific known coordinates on the Earth's surface. Your timeship will presumably come equipped with imaging tools and instruments to determine your temporal position using essentially the same methods devised by ancient astronomers to predict solstices and equinoxes, as well as a computer that can tally it all—a kind of electronic Stonehenge. But even if you can home in on an exact year and month and day with such tools, dates become more and more meaningless for locating historical events, the farther back you go. For anything earlier than a few centuries ago, you may have little or no luck finding a specific occurrence you are interested in, because the information in our history books is, at best, highly approximate.

The chronology of premodern world events that is taught in history classes was constructed during the Reformation by scholars including the Calvinist historian Joseph Justus Scaliger. It should be thought of more as a historical convention than a real historical yardstick. There are even revisionist schools of historical thought that hold that whole false centuries were inserted into history because disparate ancient timelines were erroneously laid end to end rather than mapped onto one another. A Russian mathematician named Anatoly Fomenko has argued that everything pre-1000AD is fiction. Somewhere I also read the slightly less radical claim that we are really living in the 18th century AD, not the 21st, due to some padding of the chronology around the French Carolingean period. Even without going down rabbit holes of historical conspiracy theory, this much is certainly true: Besides spacetime itself having no date markers, the cultural markers we have are worthless before the age of print (and even iffy at best, after it).

Consequently, the farther you traveled back into the past, the less likely you could ever locate a specific historical target or personage. Temporal position in antiquity might only be determined with that astronomical computer, supplementing more conventional movements of heavenly bodies with the slow precession of the Earth's axis over the millennia—1 degree every 72 years, in a cycle that repeats every 26,000 years. Over even longer periods, you could use the shifting of positions of the fixed stars that would deform the constellations in predictable ways, whichever direction you were traveling in time. But zeroing in on even a famous world-historical event like, say, "the siege of Syracuse in 212 BC"—the destination of Indiana Jones in *Indiana Jones and the Dial of Destiny*—would be all but impossible. Dial of delusion, more like.

Imagine trying to find the Crucifixion, for example—a tiny temporal target that only became famous decades after it occurred. Even if you were able to make numerous landings in different years, spoke the local tongue, did a lot of legwork in Judaea to find the target community, find the real execution site used by the Roman occupiers in more or less the right decade, right year, right season, right week, and so on, finding your quarry would be a monumental, bewilderingly complex task, even with the help of your AI robot sidekick. And that's even assuming the events as recorded in the Gospels were factual and occurred more or less as reported—the latter especially being highly, highly unlikely.

It's yet another reason why the "where are all the time travelers?" argument against time travel holds zero weight. Time travel may be doable—I obviously think it will be doable, and eventually even commonplace, and the past (and present) might be full of time travelers and time-traveling technology—but finding your way to a specific event or a specific historical personage in any historical epoch much earlier or later than that of the time traveler's home time will be challenging at best. Chrononauts will not, by and large, be visiting historic events, even the Crucifixion. And in any case, many or most of our timefaring descendants will likely have moved on to other religious figures and figureheads—we'll come back to the surprising religious motives of chrononauts at the end of this chapter.

Finding a T-rex

In a wonderful scene in Chris Marker's *La Jetée*, the man from the future appears by the side of the woman of his childhood memories in a pre-war Paris park, in front of a sequoia trunk showing historical dates associated with its rings. She asks where he comes from, and he points to a position way outside the bark—"I come from here." To paraphrase the great Jonathan Richman, some guys try to pick up girls and get called "asshole"—this never happened to that time traveler.

The beauty of this scene is to show the kind of thinking that must be applied to finding where you are, or your way back home, in time. The same way Polynesian navigators made use of a wide range of indicators—celestial, solar, zoological, oceanographic—to establish their location and find their course, future timefarers will use various natural clocks in the rhythms of planets, stars, and even life. Core samples of large trees might be part of it, not to mention radiocarbon dating of organic materials having a known time of deposition.

Certain challenges of temporal wayfinding are lessened when traveling into more remote, prehistoric periods. Targets in deep time are, for the most part, more continent-sized than island- or atoll-sized. For hitting "broad side of the barn"-type targets in Earth's geologic past, the changing periodicity of known pulsars may provide one useful gauge of elapsed time for chrononauts in a timeship, as would the frequency of supernovae throughout the galactic disk.

The latter is the method my daughters and I use on our excursions to the forests of ancient Laramidia to photograph T-rexes. (We would never dream of shooting the majestic animals, even though we know perfectly well that no "time paradox" would occur.) Donning our tachyon goggles, I spool up the Alcubierre drive, and Emily pilots our avocado-shaped vessel on a long, high arc over the galactic plane, backward-vectoring in time. From that vantage point, traveling at high speed into the past, the Milky Way looks like a forest alive with fireflies. I count flashes per minute with my stopwatch, each flash a violent stellar death, and this enables us to home in on a broad swathe of Earth time like the Cretaceous period. (Again, we're assuming a tachyon spectrum that allows us to see in the spacelike gloom of world-rewinding, and we'd have to adjust for the fact that the farther away we are from the galactic disk, the farther into its *future*, relative to our current temporal position, we are seeing.)

With crayons on typing paper she thieved from my study, May keeps a log of our exact speed and course and the duration of our retro-jaunt, so that we can (as closely as possible) mirror it later to find our way home. But to obtain that unobstructed view of the Milky Way requires venturing far from the planet, and that would not be possible without also a powerful tachyon telescope that maintains a lock on Sol—akin to a fighter jet's targeting computer keeping locked on a particular droplet of water vapor in a cloud. Without it, we would become hopelessly lost, with zero hope of finding Earth again, let alone any specific destination in its history.

Using supernovae or fixed-star positions or any other navigational method, it would be virtually impossible to find a precise event like a specific volcanic eruption or meteor impact. If you wanted to be present (or nearby—in orbit) for, say, the Chicxulub impact 66 million years ago—and by "you" I now mean an automated AI time drone with infinite patience—the only way to do so without finding and tracking the asteroid itself would be through long trial and error: thousands or millions of landings in the before and after, gradually zeroing in. (But who wants to see that impact anyway—far too sad.) Tyrannosaurids stalked ceratopsians in North America for almost 25 million years until that point, though, and that is a doable target window given our meager navigating skills, a logbook keeper just on the cusp of literacy, and a timeship cobbled together from scavenged RV parts and old electronics purchased on eBay.

At those distances, it is finding one's way back home that becomes the greater challenge. Thanks to May's log, I can try to mirror our previous journey, but there is no hope of being exact. After we return to the solar system from galactic North, it takes a while, obliquely circling our solar system, to relocate our terrestrial epoch (the Holocene/ Anthropocene). For this, we need to match the shifting shape of the constellations to their current precise configurations. From there, we return to Earth orbit and zero in further using the astronomical computer (axial precession, etc.). Lastly, we home in on technological signals (radio and television, printed matter) symptomatic of our industrial and post-industrial civilization.

We are exhausted at the end of the journey, and it is mostly because of the uncertainty and challenges of the journey itself, not fleeing from giant theropods alerted by the sound of our cameras.* The smell of warm apple pie greets us as we come through the door, and a voice from the kitchen tells us to take off our muddy boots.

Unidentified Flying Subjects

The longer humans spend as a technological species before going extinct or destroying ourselves, the easier it will be for future timeships to find their way back to their origins. Depending on how central timefaring becomes for our civilization, our descendants may erect beacons, time buoys, to facilitate the return of timeships to the right era, year, and day—powerful antennae broadcasting a frequency-varying signal that guides timeships to port. Automated timeships could also be used to create a system of temporal beacons that might make the business of navigating more distant regions of the past or future somewhat more precise and easy.

A wormhole network could perhaps be created the same way. Again, wormholes across time can be created in the vicinity of a black hole, by moving one of the "mouths"

---

* In case you weren't already aware from *Time Bandits*, the Polaroid SX-70 Land Camera is the preferred camera of time travelers. They leave something to be desired for wildlife photography, however, as they are quite noisy.

in toward the event horizon and then back out, separating them in time; but there is no way to move either of those mouths to a time earlier than the wormhole's creation. The most astounding public works project our timefaring descendents might engage in is building an aeons-spanning wormhole infrastructure using timeships to carry the mouths into the more remote past.

If their enormous energetic needs and the technical challenge of keeping the mouths open could be met, wormholes may have many advantages over timeships or could facilitate temporal exploration when used in conjunction with them. A wormhole mouth delivered via timeship to, say, the 21st century could be used as a regular gateway for more conventional air- and spacecraft, including drones. A time drone operator in the 23rd century could control and receive data from a drone in our time via a signal sent through the wormhole. The wormhole could also be used to ferry materials to and from the future in large-enough quantities necessary for infrastructure-building. Colonialist plunderers of the past could use them for large-scale theft, "resource-extraction," the same way.

My family's photo-safaris in the Cretaceous notwithstanding, the reality is surely that most adventures in time—especially, as Michael Masters points out, the initial forays, and even perhaps all adventures in deep time—will be automated. And depending on their objective, it may not be necessary for deep-timefaring drones or probes to return to the future at all. This is especially the case if wormholes prove to be difficult to create and keep open.

If harvesting non-perishable material resources, AI timebots could concentrate or stow those somewhere

where they could only be found much later, on an asteroid or obscure Jovian moon, say—a bit like pirates' buried treasure. They could similarly plant or encode information on some long-durable substrate (diamonds have recently been identified as a possibility*) either buried on Earth or, again, on an asteroid or a moon. And while it is error-prone, DNA has been shown to be a quite effective recording substrate, too.† SETI-types have speculated about ancient ET messages encoded in living genomes. Could there also, or instead, be "messages to the future"—or recorded data of some kind—written by time traveling robots in the deep past? It might be possible to intercept such messages, even if we, here, now are not the intended recipients.

These are ideas to get you thinking and imagining. The bottom line is: If your imagination is still defined by the famous light cone, you are like my neighbor's golden retriever who just had surgery, wearing a cone of … maybe not shame, but ignominious technological parochialism. The light cone is not a limit; it is an invitation. An invitation to defy our masters, bend our worldlines back on them-

---

* Jeremy Hsu, "Record-breaking diamond storage can save data for millions of years," *New Scientist*, November 27, 2024, https://www.newscientist.com/article/2457948-record-breaking-diamond-storage-can-save-data-for-millions-of-years/#Echobox=1733402253.

† Latchesar Ionkov, Bradley Settlemyer, "DNA: The Ultimate Data-Storage Solution," *Scientific American*, May 28, 2021, https://www.scientificamerican.com/article/dna-the-ultimate-data-storage-solution/.

selves, and explore the beckoning regions over the line where the sky meets the sea, and the future meets the past.

Orient Yourself

Although taking UFOs as evidence of our timefaring future would, at this point, be circular, Masters is right that aspects of the UFO phenomenon make abundant sense in time travel terms.* The imprecision and uncertainties inherent in temporal navigation offer an interesting alternative way of looking at Jacques Vallee's control-system hypothesis, for instance.

Vallee's idea is that the intelligence behind the UFO phenomenon is shaping our civilization in part by nudging our mythology. The apparent random distribution of visitations suggested to Vallee the appearance of a reinforcement schedule in a conditioning experiment, such as might be used to train animals. Except for the well-

---

* One of the most interesting cases Masters examines in his books is that of Jim Penniston, who was stationed at RAF Woodbridge in the UK during the famous Rendelsham UFO event in December, 1980. Penniston claimed to have actually touched a craft that landed in the forest. Later, under hypnosis, he claimed to remember that he was told that the occupants or controllers of the craft were from our future and that their vessels must travel in space to travel in time. He said that there was a 40–50-thousand-year limit on their range in time, lest they not be able to return. (Penniston also recalled that their purpose in venturing to our time was to procure genetic material to help with their reproductive problems in the future, which if true would support Masters' argument that future population bottlenecks due to diminishing diversity in the gene pool could lead to precisely such a situation.)

documented focus on nuclear technology,* "they" don't visit important people or leaders; they don't land on the White House Lawn. They appear haphazardly to children in a Portuguese field, to a rancher's son in Argentina, to a farmer in Brazil, to a couple on a rural New Hampshire road.

But if ufonauts are really time travelers, difficulties in locating more specific "important" targets may not allow them to appear in any other way than in a relatively random pattern, to random people. If chrononauts cannot even attempt finding specific events in the past, they may not know that much about the specific people or even places they visit. They may display a surprising ignorance. So, in other words, they might not be pretending, or putting on some Zen-koan-like surreal show for our edification, when it looks like they are lost and looking for directions, uncertain of the time (as in some cases Vallee highlights in his 1975 book *The Invisible College*†). It may be that they really are lost.

In the last several years, there has been an increasing focus on UFOs through a religious-studies lens, thanks to the work academic scholars like D. W. Pasulka (author of *American Cosmic* and *Encounters*) and Jeffrey J. Kripal (author of numerous books on the paranormal including *The Super Natural* and *How To Think Impossibly*). It is a welcome challenge to the historical military-and-defense-centered narrative on the phenomenon, as well as to the

---

* Robert Hastings, *UFOs and Nukes* ([no city]: Author, 2017).

† Jacques Vallee, *The Invisible College* (New York: E. P. Dutton & Co., 1975).

longstanding presumption that our visitors must be otherworldly astronauts gathering data. But I suspect if there is a religious dimension to the phenomenon, we should flip it—UFO witnesses could be the objects, and the visitors could be the intended subjects of the experience, again particularly if they are indeed timefarers. Basically, if a kind of ancestor-worship proves as durable and prevalent in the religions of the future as it has in the past, then some chrononauts' journeys could be religious pilgrimages. The encounters—at least, once they have found their intended target—could constitute religious experiences *for them*.

This, in fact, is the premise of one of the best time-travel novels that few people are aware of, because the author so deftly hid what he was up to. John Fowles' *A Maggot* is an epistolary novel about a motley party of 18th-century pilgrims conveying a London sex worker to a meeting with a bizarre, "Tic-Tac"-like piece of technology in a remote cave in Exmoor, in England. She is—or, will be—the mother of Ann Lee, the founder of the American Shaker movement. The reality of her experience can only be pieced together by a savvy reader of the letters that tell the story, but basically: In the distant future, adherents of this sect travel back in this machine to meet this ancestor of their faith, and give her a bewildering ride, which she interprets as Satanic (and it drives one of her companions to suicide). For the visitors, though, it is truly a religious experience.

The 18th century might be the earliest that chrononauts could hope to zero in on a specific personage, such as an ancestor or a religious founder. Although it is fun to imagine Jesus's Sermon on the Mount audience as time travelers, I have a hunch that the only people from the future who succeeded in locating him were those three kings or wise men from the "Orient" whose AI orb drone located their target and led them to that Bethlehem manger.

5.
Cosmic Circle Time:
The Universe as a Time Loop

"The hurrier I go, the behinder I get."
— 1950s slogan spuriously
attributed to Lewis Carroll

The story so far: In one way or another, fast space travel is travel into the past. A round-trip journey to Proxima Centauri via Alcubierre warp drive will return you to earth in a very different time, 2017. (Well, maybe not that different.) Thus, all interstellar voyages will be voyages into the past. Consequently, it is likely that sustained deep space exploration and expansion through our galaxy might arise as a side effect of timefaring, rather than vice versa—or more likely, the two will proceed in tandem.

The pastward vector of faster-than-light ships could prove enormously helpful in making humanity an interstellar species. Very fast, automated timeships could carry terraforming technology (such as pollution-spewing factories envisioned for terraforming Mars) to distant, uninhabited planets, and then slower settler ships (or even sublight ships) could be sent to arrive on those worlds centuries or millennia later, after they have become perfectly-tuned Edens awaiting the passengers' arrival. If those timeships ferried wormholes to their target planets, then human settlers could just step through … from one planet to another.

The uninhabited part is, of course, critical. The lie of settlers throughout European history has been to claim that already inhabited lands were empty or nearly empty so they could serve as blank canvases for colonial adventures and genocides. Critics of imperialism are thus right to question this notion of "empty space"—or, for that matter, "empty time."* In my thinking about human timefaring and expansion, my ideal model is, again, that of the ancient Polynesians whose "new worlds" were truly new, at least to our species. How could outer space not be similarly full of such islands? But I concede that we can't yet really know the truth of that emptiness.

That said, the equivalence of far spacefaring with far timefaring is one reason why we should look out in space for … our future selves. We, by which I mean the distant spacefaring descendants of our less distant timefaring descendants, could already be living on farflung worlds

---

* On this subject, see for instance Karen Barad, Rasheedah Phillips, "Tuning Into the Quantum: A Vibrational Exchange between Karen Barad and Black Quantum Futurism," YouTube, November 22, 2023, https://www.youtube.com/watch?v=5ej6y7TrfK4. In her writing, Phillips has also explored the inextricable connection between linear time reckoning and slavery: See Rasheedah Phillips, *Breaking the Master's Clock* (Chico, CA: AK Press, 2025).

throughout our portion of this spiral arm, if not our whole galaxy … or beyond. Because the farther we go out into space, the farther we go back in time. Future humanity could already be like that Polynesian superculture, occupying a vast spacetime archipelago. Who knows how far out, and how far back, it could extend.

In the 1970s, astrophysicist Ronald Bracewel talked about a "galactic club," and SETI types have debated whether we are ready to be admitted to such a club. Some speculate that our unreadiness—our xenophobia, our twitchy violence, and our fondness for nuclear weapons—has kept us excluded or quarantined. Another possibility is simpler and also more Zen-koan-like: There may be such a club, but we aren't ready to join it because *we are its past.*

Asking whether humanity is ready to join the galactic club might be like me asking if I am ready to attend my daughter May's preschool co-op. She is my most distant descendant, at this point. I gather that they play games, make crafts, and sit in a circle and sing songs. But May belongs to the future; I am just her backstory. I would probably find an experience during Circle Time as upsetting and bewildering as a wormhole-UFO abduction. Thus, I am not offended that I am not invited—it has nothing to do with my nuclear arsenal.

All this is to say, our timefaring future possibly IS our spacefaring future. And that future could already be out there. Our descendants may have evolved, in the past, in different directions on different worlds. There could be a whole Iain M. Banks post-scarcity Culture or Cultures out there, religiously preserving the innocence of their Terran Cradle.*

Some SF writers, of course, have preferred to tell a different and darker story: that human expansion through technology could actually corrupt the cosmos. In Philip Jose Farmer's Gnostic space opera *The Unreasoning Mask*, humanity's use of faster-than-light travel turns out to be a cancer on the multiverse, provoking a kind of immune response that, in previous incarnations, destroyed Earth. The idea of degenerating the body of the universe through exotic technologies like warp drives and wormholes is another neurotic idea that needs to be set aside (along with the alternative timelines, assigned walkways, and time police) if we are going to think through the problems raised in this book.

Let me try and convince you that the opposite is more likely the case. We and our space brethren might not be here at all without the time- and space-faring technologies developed in our far and farther future. Timefaring might

---

* I see some of the hands raised—you want to ask if Earth could be a terraformed outpost of an original People that originated elsewhere. Here is where evolution provides part of an answer: No. The fossil record shows a clear evolutionary trajectory to humanity via protomammals, shrews, and apes. Humans weren't just planted here, and I doubt our cosmic progenitors were shrews. Thus, if there is humanity through the galaxy, it started here. It is certainly possible, however, that life on earth was seeded here or that hominin evolution was directed by other timefaring civilizations as part of a history-gardening or very-long-term terraforming project.

be involved in the Universe's very creation. Our distant descendants might have helped make our universe, and us.

Leavening the Loaf Universe

The ancient Gnostics were fascinated with serpents, and writers in this tradition liked to rewrite what happened in the Garden of Eden, making the snake the hero instead of the villain. Rather than some evil, duplicitous tempter leading Eve and Adam astray, the serpent was exactly what he (or she, in some versions) claimed to be: a bestower of true knowledge. The Lord, in these versions, was the bad guy, jealous hoarder of what should be everybody's birthright, cruelly exiling His creations for their tasting of the Good Stuff.

Thus, it was the serpent, taking its tail in its mouth, that became one of the core symbols of the Gnostics. Ouroboros is usually seen as a kind of Mediterranean version of the Zen brushstroke circle, a nice emblem of unity and oneness in the form of a confused reptile. But the secret timefaring society to which I belong, the Order of Saint Novikov, who adopt variants of this symbol as our emblem, know that those notions are too simplistic.* Ouroboros contains a more fascinating, far more transgressive and mind-expanding truth.† That truth, smuggled out of Paradise and

---

* Igor Novikov is alive and well as of this writing, but in the distant future his mortal remains will be exhumed and found to be incorruptible, with an odor of spacetime sanctity.

† The transgressive nature of the ouroboros can be glimpsed in that distinctively American version, "The Man from Nantucket." Those who

kept in the secret custody of the alchemists for centuries, has been rediscovered by quantum physicists in the form of their famous twin-slit experiment.

Bear with me: If you shoot photons at a screen through a barrier with two narrow slits, and attach some detector that tells you which slit each one went through, they pass through the slits in straight lines and pile on the screen like little paintball pellets. If you don't attach that intermediate monitoring device, though, they behave as waves, producing an interference pattern on the screen. More amazingly, you can modify the parameters of this experiment in various ways to show that the effect is retroactive: You can decide whether or not to attach a monitoring device to the slits after the photons have already been shot out of your expensive laboratory flashlight but before you look at the screen, and they will turn out to have behaved in the way you chose.

In a laboratory it seems to be only an infinitesimal fraction of a second—maybe not enough to upset our certainties about things never ever going backwards. But listen: In a 1979 talk at the University of Maryland, the physicist John Wheeler proposed a "delayed choice"

---

know this subversive limerick know that it is about sexual self-gratification made possible by an astonishing physical endowment. (It is subversive, at least, if you are a 9- or 10-year-old boy or have one inside you.) David Gerrold, one of the minds behind the original *Star Trek*, wrote perhaps the quintessential time-travel version of this, *The Man Who Folded Himself*: A young man gifted with a time-traveling belt ends up in a world rapidly filling up with his own clones. It leads to all kinds of comic situations, culminating in a kind of homo-auto-erotic orgy with himself at slightly different ages.

version of this experiment that used light from a distant galaxy or quasar passing through the slits, to see if an Earth experimenter could dictate to that remote light how to behave, despite it having left its origin millions or billions of years in the past. It turned out that another physicist in the audience named Carroll Alley had a perfect laboratory in which to conduct a slightly more modest version of Wheeler's delayed-choice experiment: the Moon.

Science writer Amanda Gefter describes: Prior to the Apollo missions a decade earlier, Alley had persuaded NASA to have the astronauts plant mirrors on the lunar surface that could be used to reflect earthbound lasers. Neil Armstrong planted the first of these mirrors, which were to be used for making very precise measurements of the moon's varying distance during its orbit. A few light-seconds—the time it takes for a light flight to Tranquility Base and back—wasn't quite the billions of light years of an intergalactic photon journey, but it was time enough to make a quick decision how to measure some already moon-bound photons while in-flight and thereby test Wheeler's idea.*

The results Alley and his colleagues published in 1984 supported what Wheeler had imagined: The laser light seemingly "knew" how it was going to be measured already when it was shot at the moon, even if the experimenters changed their minds during those three or so seconds. To Wheeler, it seemed to open an incredible door:

---

* Amanda Gefter, "Haunted by His Brother, He Revolutionized Physics," *Nautilus*, January 10, 2014, https://nautil.us/haunted-by-his-brother-he-revolutionized-physics-234736/.

An observer on Earth could influence things way out in space—and even, by extension, in the distant past, even back to the early history of the universe. Theoretically, astronomers peering through telescopes at very distant galaxies and proto-galaxies could influence, if only very subtly, the universe's early history, through their peering.

Wheeler wrote in his journal that delayed-choice measurements of light made observation, which already occupied a central place in quantum physics, the "umbilicus of the world," and he made a drawing to represent this idea: A Sauron-like giant Eye atop one end of big U (for "Universe") gazing across the valley of cosmic history at the Big Bang.

Gefter, who immersed herself in Wheeler's journals in the library at the American Philosophical Society in Philadelphia, describes this symbol of cosmic evolution:

> As you follow the swoop of the U … time marches forward and the universe grows. Stars form and then die, spewing their carbon ashes into the emptiness of space. In a corner of the sky, some carbon lands on a rocky planet, merges into some primordial goo, grows, evolves until … an eye! The universe has created an observer and now, in an act of quantum measurement, the observer looks back and creates the universe.*

Understanding the observer-dependence of the cosmos became Wheeler's obsession for the last decades of his life.

---

* Gefter, "Haunted by His Brother."

He described such a self-causing or at least self-influencing universe as a "self-excited circuit."

In his 2008 book, *The Goldilocks Enigma*, physicist Paul Davies took Wheeler's speculations and ran with them, into realms where most scientists fear to tread: the problem of the totally improbable suitability of the universe for life. Several parameters including the exact strength of the interaction between charged particles (the "fine structure constant"), the exact magnitude of the cosmological constant that prevents the universe from collapsing into a Big Crunch, and the so-called resonance level of the carbon atom make our very existence weirdly "coincidental." These parameters were established in the first instants after the Big Bang, and had they been even an iota different, there would be no life, no intelligence, no us. It's known as the anthropic principle. The famous British astronomer Fred Hoyle was so shaken by these facts that he told a Caltech audience, "A commonsense interpretation of the facts suggests that a superintellect has monkeyed with physics, as well as with chemistry and biology, and that there are no blind forces worth speaking about in nature."*

Unless there is an Intelligent Designer who set it all in motion, there seem, to Davies, to be only two other solutions: Either there is an infinite spectrum of universes with different laws (and thus our existence is no "accident") or those anthropic parameters were somehow tuned, perhaps via those future astronomers pointing their tele-

---

* In Toomey, *The New Time Travelers*, 294.

scopes at the Big Bang. Davies wrote: "the universe has engineered its own self-awareness through quantum backward causation or some other physical mechanism yet to be discovered." * Davies also speculates that separate intelligences might converge into a vast super- or hyper-intelligence, a merging of Universe and Mind reminiscent perhaps of the Omega Point, the spiritual Singularity projected by the French Jesuit theologian Pierre Teilhard de Chardin. It's Wheeler's giant Sauron eyeball at the end of time, participating in its own creation.

More recent research in quantum computing and other physics protocols are showing that there is way more to retrocausation than an eyeball gazing across vast spatial and temporal distances. In fact, as we saw in Chapter 2, observation per se may really have been a MacGuffin in all this—a "peephole" superstitiously believed to delimit curious quantum phenomena, the same way it delimited the confusing new subject of perspective for Florentine artists and scientists like Leonardo. It probably is the case that all interactions, seen and (mostly) unseen, send causal ripples into the past. In the double-slit and related experiments that seem to show quantum weirdness, it is partly a matter of entangling the photons with the experimental apparatus itself, including that added device to show which slit the photons pass through. Photons are very sensitive to how they are being measured, and that includes a sensitivity to

---

* Paul Davies, *The Goldilocks Enigma* (Boston, MA: Houghton Mifflin Company, 2008), 250.

how they are going to be measured in their future; the same is true of electrons and everything else—even us.

It means that things are entangled with what is to come, and matter itself carries its own history backward as well as forward through the block universe. And when future intelligences alone or together develop sufficiently advanced technology, they can do more than measure and gaze—they can powerfully focus that retro-influence through the kinds of trans-temporal communication and computing methods I described, not to mention physical time travel.

Remember how little we really know of the world—our "knowledge" really amounts to assumptions; and until this historical moment, there has been little reason to assume that the world and the universe we observe could have been shaped by technologies we cannot even imagine, let alone see. The environment we see and experience and study might not only include "future technology as magic" but could even be made of it, in some sense. This offers an alternative way of thinking not only about that self-excited circuit that excited John Wheeler, but also about the possibility of reality itself as a kind of construct or, in the contemporary idiom, a simulation.

Future Technology as Nature

It would be easy for some future intellectual historian to look back on the currently popular simulation hypothesis as just one more instance of an ages-old pattern: interpreting the universe in light of the latest tech wizardry. In the Enlightenment, when the most advanced and dazzling machines were timepieces, Isaac Newton conceived of the universe as a gigantic clock.* Much of the language we still use to describe the physical world derives from the world of industrialism that ensued over the next two centuries and implicitly if not explicitly likens the universe to a gigantic steam engine. In the last decades of the 20th century, it came to be seen as a big computer—another conceptual leap often attributed to John Wheeler, who coined the phrase "it from bit" (meaning that physical reality is ultimately informational).† The related notion that the cosmos could be a big *Matrix*-like simulation, advanced by Silicon Valley game designer and entrepreneur Rizwan Virk, emerges at a historical moment when immersive, hyperrealistic virtual worlds and AI are the forefront of technology.‡

Each of these successive metaphors had, and continues to have, value in illuminating certain possibilities. Even the clockwork metaphor can still help us picture the "super-determined" gearlike articulation of a universe in which everything is entangled, across time as well as space.§ The value of the simulation hypothesis may be to help us see not simply that the universe is made of information but that that information is indeed a kind of programmable code that could be rewritten (or hacked). It also, as Virk argues,

---

* See Heinrich Päs, *The One* (New York: Basic Books, 2023).

† In the first decade of this century, MIT quantum computing researcher Seth Lloyd pushed this idea farther, arguing that the universe is a big *quantum* computer, and is programmable. See Seth Lloyd, *Programming the Universe* (New York: Vintage Books, 2006).

‡ Rizwan Virk, *The Simulation Hypothesis* (xxxx: Bayview Books, 2019).

§ See Silberstein et al., *Beyond the Dynamical Universe*; Päs, *The One*.

resembles ages-old mystical insights about underlying layers of reality ordinarily hidden from view, as well as provides new ways of thinking about paranormal phenomena like UFOs that might reflect technology tapping into those deeper, code-like cosmic laws. Arthur C. Clarke's famous formulation that any sufficiently advanced technology would appear to us as magic would, in Virk's view, need to be extrapolated upon: If a sufficiently advanced technology will appear as magic, an even more advanced technology might simply appear as ordinary reality, or nature itself.

But every metaphor also carries baggage and is destined to be superseded by the next technological revolution. The glitches that Virk and other writers see as possible clues to the imperfect rendering of our simulated reality have many other possible explanations. The Mandela effect, commonly cited in this connection, is more easily explained through quirks of individual and collective memory.* The simulationist notion that many or most of the people we inhabit the world with could be NPCs, perhaps appealing to a gamer, has morally odious implications if taken literally. And in any case, the idea that advanced beings would build and perhaps escape into

---

* Shared mis-memories were named the Mandela effect because many non-South Africans were surprised to learn that Nelson Mandela died in 2013, as they remembered him dying in jail in the 1980s—hence the idea that maybe they are remembering a different timeline. They are more likely confusing/conflating Mandela with Steven Biko, a dissident who was killed in police custody in 1977 and who was the subject of a 1980 Peter Gabriel hit single ("Biko") and a 1987 film, *Cry Freedom*, starring Denzel Washington.

game-like simulations—or become Archons keeping sentient beings imprisoned in them—strikes me as less interesting and plausible than something like Davies' future superintelligence actively co-creating or cultivating its real universal home via various forms of time travel and macro-retrocausation. In other words, history gardening. Maybe we could even call it reality gardening. Imminent technologies of time manipulation like quantum computing could be a sort of tiny prelude or warmup for such an ultimate cosmic anthropic horticulture.

One as-yet-underappreciated implication of various retrocausal or otherwise future-determinist models like the relational block world is that they open the way to seeing how, on the most fundamental level, the physical world must be made of causal loops. I have speculated elsewhere, riffing on some of Davies' ideas, that if organic molecules scale quantum time loops up, it could help explain life's arising from inorganic matter.* And my books on precognition have argued that brains and cultures scale them up further, as tesseracts that carry information backward from the future, within the individual's lifespan and within the long life of a society or civilization, respectively.† Could the genome work similarly for species, or even for life as a whole?

Those quantum servers in *The Peripheral*, connected to 3D printers in the past, could provide one technological

---

* Eric Wargo, "The Wyrd of the Early Earth: Cellular Pre-sense in the Primordial Soup," *The Nightshirt*, May 22, 2016, https://www.thenightshirt.com/?p=3840.

† See Wargo, *From Nowhere*.

metaphor for how a kind of retro-programming of life could occur in such a picture. Some recent research is suggesting that, like the microtubules of the cytoskeleton that may give neurons their precognitive power, DNA too could have quantum-computing properties.* If so, then perhaps the double helix in the cells of all Earth life is a bit like that quantum server: tied to "peripherals" (organisms) that 3D-print its instructions via transcription and protein synthesis, potentially directed or at least nudged by commands issued in the future and not just prior selection.

A visual metaphor for a world made of time loops would be of a body constituted of cells doing little functional ("machinelike" if you like) things within larger assemblages of organs and organisms and collectives. Life, mind, and culture all seem to be built of cell-like time-looping formations in spacetime whose intricate inter-relationships work a little like interlocked gears to drive the emergence of coherent orderly structures out of the material chaos. Verbs like "drive" are clumsy approximations in an adynamical block universe, but thus far we really have no better language to talk about this. While any particular level of organization may give the rough appearance of "producing" in a rational, linear, machine-like way, each of those cells or gears is just acting like that time traveler in Michio Kaku's "information paradox" thought experiment—delivering to his younger self the

---

* Elisabeth Rieper, Janet Anders, Vlatko Vedral, "Quantum entanglement between the electron clouds of nucleic acids in DNA," arXiv:1006.4053v2, https://arxiv.org/abs/1006.4053.

plans for a time machine, something coherent that was never made anywhere or anywhen, just "found."*

The engine of emergence in this view would not be some thermodynamic linear process but simply the dazzlingly simple (so simple it is hard to conceive) non-contradiction of events in a self-consistent universe where causes propagate in both directions—Novikov's "force" erupting from the micro to the macro. Macro-retro-causation in the form of time travel and related technologies may in turn act as a control system on lower levels of causation, lower-order systems, bringing space, time, and matter into alignment with some cosmic anthropic telos rather than destroying or degrading the fabric of the universe, as Farmer imagined. In this way, timefaring technology, or even the wormhole trails it leaves through the Minkowski block, could even be the yeast that has made that block (or loaf) fit for life (and mind, and culture).

We veer into wild speculation, of course. But again, the point of this book is to broaden our imaginations. These speculations don't all have to be right. ("We don't have time for being right about everything," as Jeffrey Kripal acutely notes—life's short.†) But how many yet-unexplained anomalies (from the standpoint of our resolutely

---

* I really wonder whether the Novikov self-consistency "force," in a universe where particles are constantly looping in time and not causing paradox, could be as fundamental as the other forces of nature, or perhaps could be the still-elusive bridge between the quantum world and gravity.

† Jeffrey Kripal, *The Superhumanities* (Chicago, IL: The University of Chicago Press, 2022), 30.

linear-causal Newtonian viewpoint) might reflect higher-order cultural and technological time loops, as well as interventions of our human or posthuman descendants in their histories? I expect the makers of our world are not outside it, looking in. If the laws of nature themselves are programmable, it could be Wheeler's God-Eye and Davies' Hypermind at the End of Time doing the programming.

And really, that Hypermind doesn't need to be at the End of Time. While timefaring technology may arise in our future as the outcome of a mostly linear story of civilizational progress, the sufficiently advanced civilizations to which we will give rise need not be confined to inhabiting that future; they may permeate the whole history of the universe, for the reasons we explored at the beginning of this chapter. It means those history gardeners can and will have been active cultivators of cosmic history and even physical law from the get-go, the leavening in the cosmic loaf. Again, our future is in the past.

Wheeler's God-Eye could really be a God-*Hand*, and it could be right there at the start of things, setting things in motion, tinkering with the parameters, getting the balances just right.

6.
A Sound of One Hand Clapping:
Meeting Your Older/Younger Self

You've probably figured out that the futurist argument I have been making in these chapters is smuggling a philosophical argument inside it. Physical time travel will likely be our reality, sooner or later—I do believe that—but even before that happens, this knotty subject can also be a powerful philosophical tool for thinking about the weird, "impossible" reality of a single, actual world.

The tendency of philosophy, psychology, and physics (not to mention fiction) over the past century, and especially in the last few decades, has been toward radical questioning of an objective universe outside our perceptions. Quantum physics, as we saw, described what seemed like a basic uncertainty at the heart of nature. Attempts to make sense of this uncertainty sent different schools of physicists in different directions, but always those directions only got us farther and farther from any notion of a single coherent reality "out there." In the thinking of loop-quantum-gravity theorist Carlo Rovelli, for instance, there is no single reality but an unending multiplicity of realities, an infinite plurality of observers collapsing their own wavefunctions.* It is the exact horror that Eugene Wigner imagined with his thought experiment about Schrödinger's cat's owner and the owner's friend (and friends' friends, and so on).

This decentering of reality has been anticipated and echoed across all disciplines and domains of Western knowledge. The social sciences, such as my own field of anthropology, came to view reality as socially constructed; and academic postmodernism called all assumptions about self and truth into question. Two decades before the simulation hypothesis came into vogue, philosopher Slavoj Žižek was already arguing how computer-mediated virtual reality helps us see that reality itself is virtual.† Lately, UC Irvine cognitive scientist David Hoffman has made a case that we don't perceive reality at all but only "icons" on a kind of subjective computer screen—a neuroscience version of the simulation hypothesis.‡

In short, the psychological and cultural constructs of modernity and postmodernity have eroded our ability to think about anything solidly *actual*; we cannot really

---

* See Gefter, *Trespassing*.

† Slavoj Žižek, "From Virtual Reality to the Virtualization of Reality," in T. Druckrey (ed.), *Electronic Culture: Technology and Visual Representation* (New York: Aperture, 1996).

‡ Donald Hoffman, *The Case Against Reality* (London: Penguin, 2020).

fathom a place that is not supported by some no-place, or a history that is not supported by multiple counterfactual stories.* In this environment, the actual has become the greatest heresy, the biggest impossibility.

But there is a saying in Zen Buddhism: Before studying Zen, mountains are mountains; while studying Zen, mountains are not mountains; after studying Zen (that is, after enlightenment), mountains are mountains.

Like the return to mountains being mountains, retrocausal or relational block-world frameworks could return us to a singular, actual, real world—not, however, as a throwback to some naïve realism, the sort of thing Niels Bohr faulted in Einstein, who maintained a faith in a world existing independently of the observer. Really, these frameworks herald a new kind of "constructivism" in which the observed world, the world prior to our inter- or intra-action with it, *was* already there, already real; yet amazingly its reality was constituted in part by that future inter- or intra-action.†

When we face this possibility—and converging lines of evidence from multiple fields are hinting it is true—then those virtualities and multiplicities and multiverses all begin to look like avoidance tactics or defense mechanisms not just against "things going backwards" but also against that scary but (I suggest) really liberating actuality of things. There always was a real world, even if we could never and can never see or know it clearly; our actions

* See Wargo, "Edges of Forever."

† Karen Barad uses the term "intra-act" in her extraordinary book, *Meeting the Universe Halfway*.

moment to moment, including our observations—but really, all the actions of everything, sentient or not—co-constructed the world we inhabit. The world at any given instant is a compromise—maybe the *only possible* compromise—between influences "pushing" from the past and another realm of influences pushing back from the Not Yet.

The reality of a single self-consistent universe, a "this world" that is unitary, that possesses a single history, that exists even when we are not looking at it, and that we need to take care of because it is "all there is," is the surreal reality that Homo Chrononauticus will have evolved to embrace and even love. Yet for many people today, addicted to those virtualities and multiplicities, "all there is" sounds imprisoning or at least depressing, like the 1969 Peggy Lee hit about existential disappointment, "Is that all there is?":

> When I was twelve years old, my daddy took me to the circus, the greatest show on Earth. There were clowns and elephants, dancing bears, and a beautiful lady in pink tights flew high above our heads.
>
> And as I sat there watching, I had the feeling that something was missing—I don't know what. But when it was over, I said to myself, "Is that all there is to the circus?"*

---

* Peggy Lee, "Is That All There Is?" (written by Jerry Leiber and Mike Stoller), from the album *Is That All There Is?* (Hollywood, CA: Capitol, 1969).

To feel disenchanted by the singularity of the actual is, objectively, strange: "This," the universe, especially as it has been disclosed (and yes, constructed) by centuries of scientific exploration, is quite impossible huge and complex—it may as well be infinitely vast and comprises laws and phenomena we can barely yet fathom. The immensity and complexity grow exponentially, the more we investigate and learn. In the infinitely vast universe, is it somehow more inspiring to imagine there are infinite infinitely vast universes? How many infinities do we need to be satisfied? In which of the infinitely vast universes in the multiverse can we find the meaning or satisfaction that might be eluding us here and now?

People make and find meaning in approximately 8.2 billion different ways, but the reality of time loops, and that "force" of non-contradiction in a single, self-consistent universe, supports the Zen path I have always found most helpful in my own meaning-making. I believe it can help us think anew about the fact of our being right here, right now, in a single place within a singular world. When meditated upon, it can lead to an epiphany about the beauty and profundity of things just as they are, alchemically transmuting Peggy Lee's existential disappointment into the gold of what the Buddhist writers called "suchness."

The Self-Encounter

The particular scenario I have in mind, that could move us to an appreciation of our radical actuality, is that very encounter that scared so many physicists off, even when the math and science showed time travel to be possible: the self-encounter across time, meeting one's younger or older self. It was so unsettling that it deflected them down various blind alleys such as those multiverses, or caused them to immediately leap to the possibility of *killing* that other version of themselves—again, like a confused cuttlefish attacking its reflection in a mirror.

Conflict makes great drama, so if nothing else, those violent thought experiments are certainly an entertaining sidelight to dry, math-laden theoretical physics papers, just as they make entertaining premises in sci-fi paperbacks. In his story, "The Seventh Voyage of Ijon Tichy," Stanislaw Lem has his hapless astronaut-hero encounter multiple versions of himself after his spacecraft is damaged from being hit by a time vortex. It is a kind of Keystone Cops situation in which there is just one "cop," Tichy himself at various temporal displacements, teeming up to try and repair a busted rudder on his rocket, but getting bogged down in bickering and acrimony. More seriously, the Netflix series *Dark* traces the knotty time-traveling long self of a character initially named Jonas. At one point, when Jonas is only beginning to glimpse the temporal weirdness going on in his world, he meets with a Stranger on a bench at his father's grave site—it is in fact his fortysomething midlife incarnation who does not divulge his identity but enigmatically counsels his teenage self that "life is a labyrinth" and that he should always be asking himself if he is making the right decisions.

It needs to be said that such self-encounters will likely be extreme rarities in our actual timefaring future, for various reasons having little directly to do with alleged paradoxes or causality violations. First, the overwhelming majority of place-times in the universe did not include the time traveler—we are tiny worldline-threads in that vastness—so such encounters will be unlikely to occur by accident. And perhaps more importantly, time travel technology will be (or is) some of the most expensive, powerful, and potentially disruptive (or destructive) technology imaginable. As Lem puts it in another Ijon Tichy story, "Chronomotion … is a colossal enterprise, requiring tremendous investments, expenditures, planning …"[*] So an individual chrononaut's visit to her childhood, for instance, will not be on the Time Program's mission list.

That could change with the creation of a timefaring infrastructure or an "always open" wormhole network, if the technical and power challenges could somehow be solved. But even if and when timefaring does become easy, a basic understanding of the self-consistency laws of the universe will also tend to filter out intentions to self-meet. Some version of the following thought-train, for instance, may dissuade you: You wake up in the morning with the impulse to travel back in time and visit your younger self, but … you don't remember a visit from your older self, when you were younger. Knowing that you can't change history, you abandon the idea. (That's an easy case, requiring no causal intercession from Saint Novikov.) But perhaps you find this "obedience to what was" frustrating, and your free will kicks in: "Fuck it, I'm going anyhow."

---

[*] Stanislaw Lem, *The Star Diaries* (Orlando, FL: Harcourt, Inc., 1976), 129.

You figure that maybe you'll find that you're unable to reach your target, somewhat like in Ted Chiang's "The Merchant and the Alchemist's Gate," or maybe you'll have your encounter, and it will turn out that something happened later to give Older You amnesia for the incident. … And *this* train of thought leads you again to rethink your journey: Why would I go if my visit will have no effect on my life now? What effect am I trying to achieve anyway? Why try? There's no sense in *trying* across time. It either happened or it didn't. It's easier (in most cases) to just fast-forward to it didn't. (Again, Blessed Saint Novikov, you are off the hook.)

Ceasing to believe in the possibility of changing the past is part of the necessary backstory of being in any position to utilize time-travel technology. Consequently, most of our chrono-adapted descendants will not even be interested in interacting with their past selves but will opt instead to visit histories remote from their own, to achieve unanticipated outcomes. If they do visit their own pasts, they may do so stealthily, as an observer, to learn new things they didn't know—perhaps things that will alleviate doubts or regrets, as in Chiang's tale—but not with the intention of an actual self-encounter.

In most cases.

But there are, will be, exceptions: What if you *did* remember a visit from a strangely you-looking older person or from someone who introduced themselves as your older self? What then? And what, in such a time-looping life, would Older You then say to Younger You upon your return, circling back, all those years or decades later?

***

In its third season, *Dark* became a story about multiple timelines: We find out that Jonas eventually became Adam, the elderly founder of a faction called Sic Mundus (short for the hermetic motto, *Sic Mundus Creatus Est*, "Thus the World Was Created"), living in 1921 and waging a time war with a rival faction led by his middle-aged iteration, the Stranger. This creates a nice dramatic reason why the Stranger would have been so cagey in his first meeting with his younger self—the history of this world is *not* self-consistent, so he doesn't know, simply from memory, what his younger self, Jonas, will do, and everything really is uncertain. It makes sense to give Jonas vague advice about making the right choices but not to divulge too much.

Navigating time's labyrinth is staggeringly difficult when keeping to one block universe, so I don't necessarily blame *Dark*'s creators, director Baran bo Odar and writer Jantje Friese, for copping out with the "many worlds" device in much the same way William Gibson did in *The Peripheral* with his "stubs." But our task in this chapter is to push through the difficulties and imagine how a transtemporal self-encounter might plausibly unfold within a single, self-consistent universe, a single history. Real secret societies of timefarers might not only foster such occurrences but might do much more with them than *Dark*'s writers did, indeed making them a kind of *initiation*. Initiation is really built in to the encounter's basic, riddle-like structure.

Here's one way it could go (I imagine there are others).

Younger You meets an older version of yourself who already lived through this encounter and remembers it. To lend this meeting even greater seriousness—it is a meeting about your destiny, after all—imagine it taking place by a parent's or other ancestor's grave, as in *Dark*. Because Older You understands time travel, you realize that what Younger You did in life after that meeting, they did anyway, so giving advice like "make sure you are making the right choices" is really a kind of empty platitude. What might you say instead?

First, of course, recall as best you can: What *did* Older You say then? In fact, in reality, the encounter was hugely important, an initiation, so at least something of it will have stuck in memory. But memory is never exact, and that imprecision is in fact the saving element in all this. As you imagine saying something that you already did say, you (and now I mean you, the reader) begin to balk: You are imagining that the encounter would be "going through motions" or doing something already done, or giving advice for something you already did. It's not logically inconsistent, but it chafes against our demand for freedom. No one wants to be merely obedient to what was, to feel as if they are in a play. But that fickleness and fleetingness of memory makes your recall uncertain and untrustworthy—it is the looseness or "play" (in the untightened mechanism sense) needed to alleviate that feeling of play-acting.

So, really, the question you must ask yourself is: What do you *want* to say? What would you say, naturally and spontaneously, to your younger self?

In a self-consistent universe, what you want to say and what you did say should match up: You will say it not only because you did say it, but because you really want to say it.

If you are like me, you might have trouble imagining such a congruence, because most of us don't entirely live the life we thought we wanted to live when we were younger, or that we would wish our Older Self could have counseled us to live. We live in and with regrets. James Gleick was not wrong that time travel can hinge on regret.*
But in a self-consistent universe where you really understand that you can't change the past but also that you don't really *know* the past, then that regret can be transmuted into something else.

Inner Conflict

One way of understanding regret is as a kind of "inner conflict" strung out across time, sequentially. I have argued in my other books for a temporal rather than spatial framing of the Freudian unconscious with its various tiers and components. The unconscious is future consciousness as it interacts with present consciousness; and Id, Ego, and Superego are all taking turns, just as the different priorities of a life fully lived tend to succeed and supplant each other in time.† In other words, it is not that we have multiple parallel inner agendas in simultaneous conflict, but that we go through life perpetually thwarting or undercutting our older selves and betraying our younger selves. This can be seen in things as simple as procrastination or keeping to a health or financial regimen, on up to the big existential

---

* Gleick, *Time Travel.*

† See Wargo, *Time Loops.*

crises and neuroses that make our lives painful and that structure our biographies.

The conflicting interests of You Now with You Yesterday and You Tomorrow may have a lot to do with the seeming codedness of messages in dreams, which I think are often if not always precognitive. In fact, the enigmatic nature of the self-encounter across time—for instance, Coop's message to his younger self via bookshelf-tesseract in *Interstellar*—could serve as a model of how information refluxes back in time in a tesseract brain. A dream symbol must have a dual (or more) meaning: It must have an immediate sense or sensibility, which impels some kind of action that feels freely willed, and then a final or teleological meaning, a "real" meaning, that only becomes apparent once the symbolized event comes to pass.

This dual (or more) structure to the symbol orients the individual toward challenges that are withstood: Not away from calamitous or terrible outcomes, nor headlong into them, but just around or across their periphery. Consequently, the core "message" of dreams, whether they are about trivial events like sinks backing up or plane and car crashes, can often be boiled down to: "but I survived" (or perhaps "I'm still standing.") The same applies to the self-encounter as a time-travel initiation.

Dark's creators were using the time-travel trope to tell a story of inner conflict, through the vehicle of a war between Jonas and his selves at different ages. But one thing I think is certain about the self-encounter if it occurs within a single consistent timeline is that Older You does *not* withhold their identity, as the Stranger does: "In case you are doubting, I am You, in X years." This is crucial.

From Younger You's standpoint, this encounter now contains the second most positive and empowering piece of news that a mortal being could hope for: You survived … at least until the age of Older You at the time of the meeting. (I think we need to imagine the self-encounter as spanning a few decades of the timefarer's long self.)

One thing that keeping to a single timeline both allows and probably even demands is that the core metamessage of the self-encounter becomes that of survival. As long as Younger You trusts that Older You is really who they say they are, and as long as you understand the rules governing a self-consistent universe (that your future life MUST come back around to this point, X years later), then this meeting has a somewhat similar effect on your psyche as "dying before dying" in the psychedelic mystery rites of the ancients: An assurance of power and semi-permanence. Nothing you do in those two or three decades can destroy you.*

What a gift! It is the gift that members of secret societies of timefarers will (do) give to their younger selves and that makes them, in effect, indestructible during their more robust years of youth and middle age. They achieve amazing things during this time, because they are fearless. They can't die, because they didn't die.

---

* I said this message was "second most positive"—the message of eternal life beyond death, reunion with the departed, and so on, would be greater, of course, but that may not be in the power of even Master Timefarers to grant.

Again, I suspect that this effortless logic of non-paradox, the Novikov force, may be the real Force underlying the universe and its solidity—below even the elementary forces governing particle interactions in nature. The Secret jealously guarded by timefaring secret societies is knowledge of this Force, which truly is all around us and we do not see it. In their secret fortresses on hidden moons, they conduct other kinds of experiments where they pull Jinn out of empty history, tinkering with the first principles of Creation. But in a way, via these initiatory self-encounters they turn *themselves* into Jinn, because these encounters have a telescoping effect on their lives.

However much or little Older You says on that bench, Younger You sees the signs of your future success, and they naturally rush to fulfill it, feeling as if the world will open up to their goals and objectives without obstacle or impediment. Because confidence is indeed a key that opens many locked doors in life, the effect proves generally self-fulfilling. It is a more positive version of what happens to Macbeth after listening to the Weird Sisters' prophecies—he rushes to fulfill them. Unlike Macbeth, however, your confidence cannot fail—or at least, it cannot prove fatal. There is no doubt that Older You has lived a weird life—it goes with the time-travel territory—but whatever challenges you faced, they did not defeat you (unlike poor uncurious, literal-minded Macbeth). Per the old Nietzsche quote, the obstacles did not kill you so they made you stronger.

By the same token, however, there may be no reason for Older You to be too explicit about your life and travails to come, so giving advice like "do this and do that" (or "always make the right choices") may not be part of the encounter, because it won't change anything and would only cause the stubborn free-will of Younger You to kick in.

As I envision it, the older self presents themselves as who they are, but does not give advice or guidance. It may be enough to just sit in silence after introductions are made, not unlike the "flower sermon" of Zen Buddhist legend, where the Buddha just held up a flower, saying nothing, and one guy in the audience, Kashyapa, got the message, and smiled.

But here's the thing: The initiatory self-encounter on the cemetery bench turns out to be a double initiation, because it is an initiation for Older You as well—a very different kind of initiation. This is the part that you didn't understand, going in.

Congruence

As the appointed encounter with Younger You looms—it has been arranged and prepared by your hooded brethren in the Mothersaucer—you feel intensely conflicted. Your vague memory of that encounter, all those years ago, was that after the initial awkwardness, Older You was actually happy, emotionally overwhelmed, maybe fighting back tears … yet now, having become Older You, all you feel going into the meeting is dread. The weeks and months leading up to this scheduled meeting have been some of the worst in your life, in fact, where you have become paralyzed by anxiety, even succumbing to the akinetic mutism of more time-confused people when they first grapple with the possible nonexistence of free will. You thought you had put such confusions behind you back at Time Travel Academy, but now all is uncertainty and stress: You can't resist the temptation of replaying memories of the encounter from decades ago, and your trepidation feels highly incongruent with what you remember.

Knowing that everything beyond the meeting with your younger self will be uncertain, open and unknown (rather like the gloom of the spacelike), you approach the meeting with actual fear. This encounter marks the horizon of that temporary immortality that originally had empowered and emboldened you. In fact, your dread is the flip side of the courage the encounter had imparted to Younger You, and it feels like a debt that suddenly needs to be repaid with an unfair burden of interest. Since your trepidation doesn't match the happiness you remember yourself showing on that bench, you have come to distrust that memory, and you start to wonder if those tears of joy had really been tears of something else.

But here's what Older You will discover, sitting next to Younger You: The truth of the self-encounter across time is that the Younger You is the teacher of the Older You, not the other way around. The younger is the stronger of the two of you, giving something of that strength to the older, and in a way that the older only realizes in that very perplexing moment. Sitting next to slightly bewildered Younger You, Older You's dread will turn into a kind of love and wonder. *This really happened. This was real.*

By extension from that fact, it's *all* real, all singular and permanent in the Block Universe. It's a glorious, wonderful thought, even if individual timefarers' world-lines must eventually come to ends, like those of all mortal beings. So, all that heightened mortality fear Older You had been experiencing is replaced by a kind of enlightenment, an accepting of impermanence within permanence or permanence within impermanence. Those remembered tears now make sense.

Mainstream psychologists are increasingly interested in methods to foster greater connection to our future and past selves, as an approach to building mental health. Earlier I mentioned neuroscientist Julia Mossbridge, who is doing this with her efforts called TILT (The Institute for Love and Time). An active precognitive dreamwork practice can strengthen self-connection.* Many magickal or occult practices are accomplishing the same kind of thing. So can

---

* Precognitive dreaming and dreamwork, as a form of informational time travel, provides a model for the kind of epiphany time travelers would experience in a self-encounter. This is especially true for dreamworkers who keep long-term dream records. Opening a very old dream journal from decades in the past and reading, right there on the page, a description of a big event that just happened or is right now happening in your life feels like your younger self was unknowingly sending you a big and important message of reassurance.

Although the idea of sending deliberate messages to one's younger self recurs in thinking and writing about precognition and the long self, the necessary obliquity of such messages, the fact that they are bound to be misunderstood at the time we have the dream, makes the content of such messages perhaps less important than the simple fact that they exist—that there *is* a long self, that you *do* continue through time.

writing or an art practice like photography or painting.[*] Ancestral religions, of course, build something like love and congruence (and many other things) over greater spans of time than the individual's biography.

Whether initiatory self-encounters unfold exactly as I have imagined here or not, the reality of time travel will tend to foster that kind of congruence over the span of life—a general tendency to always remember our younger selves and not betray them, a tendency to know we have older selves and not throw *them* under various buses. Timefarers will tend to bring peace and joy to their selves at other ages, and an almost supernatural encouragement. Elderly selves will not wage time wars with their rebellious younger selves but will honor them and their remembered intentions. And middle-aged selves will not mysteriously appear on benches next to their teenage selves without revealing who they are. And no selves, at any age, will be taking pot-shots at themselves through wormholes.

Occasional temporal Jinn notwithstanding, it is mainly when our older and younger selves are in such a state of congruence that they will actually meet, whether it occurs on a cemetery bench, on a city street, or on a high mountain peak. The congruence will cause the meeting, and the meeting will cause the congruence, in a loop.

As the classic Zen koan goes, "What is the sound of one hand clapping?" The answer is You meeting You across decades, and shaking hands. *That* is the point of the

---

[*] My own practice involves keeping fetish-like souvenirs of my past and future, in the form of Polaroids, dream journals, and significant objects, giving me the appearance, sometimes, of a hoarder.

encounter on the bench. And it is really the point of becoming timefaring.

Acknowledgments

This book originated as a paper, "Where the UFO Was, There Will I Be," presented at the Esalen Center for Theory and Research in Big Sur, California in January, 2019. Versions of some of these chapters appeared on Medium and in a few cases were based on posts that originally appeared on my blog, *The Nightshirt*. My thanks to readers who gave me feedback on those articles. I also thank my wife, Laura, for her encouragement, our daughter May for her patience, and our daughter Emily for her assistance with the timeship animations. And huge thanks to David Metcalfe for art and saucers.